Britain's f[...]
atomic tests i[...]

THE SECRET OF EMU FIELD

ELIZABETH TYNAN

NEWSOUTH

A NewSouth book

Published by
NewSouth Publishing
University of New South Wales Press Ltd
University of New South Wales
Sydney NSW 2052
AUSTRALIA
newsouthpublishing.com

© Elizabeth Tynan 2022
First published 2022

10 9 8 7 6 5 4 3 2 1

This book is copyright. Apart from any fair dealing for the purpose of private study, research, criticism or review, as permitted under the *Copyright Act*, no part of this book may be reproduced by any process without written permission. Inquiries should be addressed to the publisher.

A catalogue record for this book is available from the National Library of Australia

ISBN   9781742236957 (paperback)
        9781742238388 (ebook)
        9781742239286 (ePDF)

Internal design Josephine Pajor-Markus
Cover design Mika Tabata
Cover image Atomic bomb test, Emu Field, Great Victoria Desert, South Australia, 15 October 1953, United Press Photo, National Library of Australia

All reasonable efforts were taken to obtain permission to use copyright material reproduced in this book, but in some cases copyright could not be traced. The author welcomes information in this regard.

FOREWORD

The Secret of Emu Field is a much-needed and valuable addition to the history of the British nuclear tests in Australia. Until now, there has been no book that focuses specifically on Emu, although there has been a plethora of books (both non-fiction and fiction), documentaries, screenplays and theatre scripts on Maralinga. There has even been a satirical television mini-series on the Maralinga atomic tests, much to the horror of the Maralinga Traditional Owners.

Little was, and is, publicly known about Emu.

Between 1953 and the Royal Commission into British Nuclear Tests in Australia in 1984, few people in Australia even knew there was a place called Emu Field. They had probably heard there had been atomic tests at Maralinga in the 1950s but not at a place called Emu.

Publicity about the effect of the 'black mist' from the Totem I atomic blast on Yami Lester and his Yankunytjatjara community at Wallatinna only started emerging in 1980, which came as a huge surprise and was one of the factors that gave rise to the 1984 Royal Commission.

The Traditional Owners of the Maralinga Lands accepted freehold title over their country on 17 December 1984 while the Royal Commission was in progress, apart from the test sites at Emu and Maralinga, which were handed back in 1991 and 2009, respectively.

I had the great privilege of acting as junior counsel to Geoff Eames QC for all Aboriginal interests before that Royal Commission. As a result of the Royal Commission, we were able to negotiate compensation for the Wallatinna mob and every other Aboriginal person we could identify as having been affected by any of the nuclear tests in Australia.

The Traditional Owners then negotiated the rehabilitation of the nuclear test sites at Maralinga and Emu between 1986 and 1993. The major rehabilitation was completed in 2000. Most of that rehabilitation work was at Maralinga, primarily because that was the location of most of the long-lasting plutonium-239 fragments, a legacy of the Vixen B minor trials. Again, the spotlight was on Maralinga and not Emu.

The developing story of the impact of the black mist from the Totem I blast on Aboriginal people has been, until this book, the only spotlight cast on the Emu tests. That story would not have taken shape or resulted in the acceptance of the Aboriginal accounts of the black mist and the negotiation of compensation but for the Royal Commission and the persistence of the late Yami Lester.

Western Desert Aboriginal communities and Australian society owe a great debt of gratitude to Mr Lester for his dogged but charming determination to bring this awful chapter of British and Australian history, which is so well depicted in this book, to light.

Elizabeth Tynan has treated the Aboriginal legacy of the Emu tests with rigour and sensitivity.

Evidence of the effects of Totem I on Aboriginal people at Wallatinna and Mintabie emerged during the 1984 Royal Commission and the likely health effects are well presented and analysed by the author. As was submitted on behalf of all Aboriginal interests in the Royal Commission, the full health effects on Aboriginal people can never be known, given the absence of health services and records for Aboriginal people in the region prior to Totem I.

Differences of opinion, unproven assertions and uncertainties about these matters remain. The author has identified these and considered them objectively and sympathetically.

Much of this well-researched and well-structured book is new material. The history of the negotiations between Britain and Australia over atomic tests, the aspirations of the Australian Government to have a significant role and the extent to which Australia was sidelined and fobbed off are adroitly assembled here, providing a rich context to the two atomic tests and five minor trials performed at Emu.

Most Australians would not have known that Australia saw Emu as a huge opportunity to exploit its uranium resources at Rum Jungle. Neither were we aware of the full extent of Professor Ernest Titterton's ready seduction by the British from, ostensibly, Australian scientific representative to British stooge. The English-born Titterton was made to feel an important cog in British weapons development.

Further, as Elizabeth Tynan points out, the lessons learned at Emu were irrelevant to British nuclear weapons development even before the next atomic tests commenced on the Monte Bello Islands two and a half years later. Emu has thus remained a footnote to the British nuclear test program but a disaster for Aboriginal people of the Western Desert.

The author has given the Emu tests a shape, a colour and a context which, for the first time, pay proper respect to the test sites near the Emu claypan, the surrounding country and the Aboriginal Traditional Owners—the Pitjantjatjara, Yankunytjatjara and Antikirinja traditional and secular custodians of that country.

Today the Emu test sites are part of the Maralinga Lands but also remain part of the Commonwealth Defence Department's Woomera Prohibited Area. Access to the Emu sites requires a permit from the Maralinga Traditional Owners and from the Defence Department. Travel by road between Emu and Maralinga is prohibited to all but the Traditional Owners and Defence personnel. This is because the Maralinga Traditional Owners are now responsible for the security of the former atomic test sites, a responsibility they take extremely seriously.

This book finally and properly puts Emu on the map and gives Emu its true but vexed place in Australian and British history.

Andrew Collett AM

Dedicated to everyone
who was harmed by
British atomic tests
in Australia

CONTENTS

EMU FIELD OPERATION TOTEM 1953 MAP	XII
ABBREVIATIONS	XIV
MEASUREMENTS	VX
PROLOGUE	1
1 FINDING X200	17
2 WHY EMU?	42
3 TOWARDS A BRITISH 'AUSTERITY BOMB'	65
4 SOUND AND FURY AT EMU	92
5 THE UNKNOWABLE BLACK MIST	117
6 SECRETS AND SAFETY LIES	142
7 FLYING THROUGH THE CLOUDS	164
8 THE PEOPLE'S WITNESS	188
9 AUSTRALIA'S ATOMIC BARGAINING CHIP	212
10 MARIE CELESTE	237
11 THE FORGETTING OF EMU	257
APPENDIX: BRITISH ATOMIC TESTS IN AUSTRALIA	280
GLOSSARY	282
NOTES	287
BIBLIOGRAPHY	318
ACKNOWLEDGMENTS	337
INDEX	344

MAP

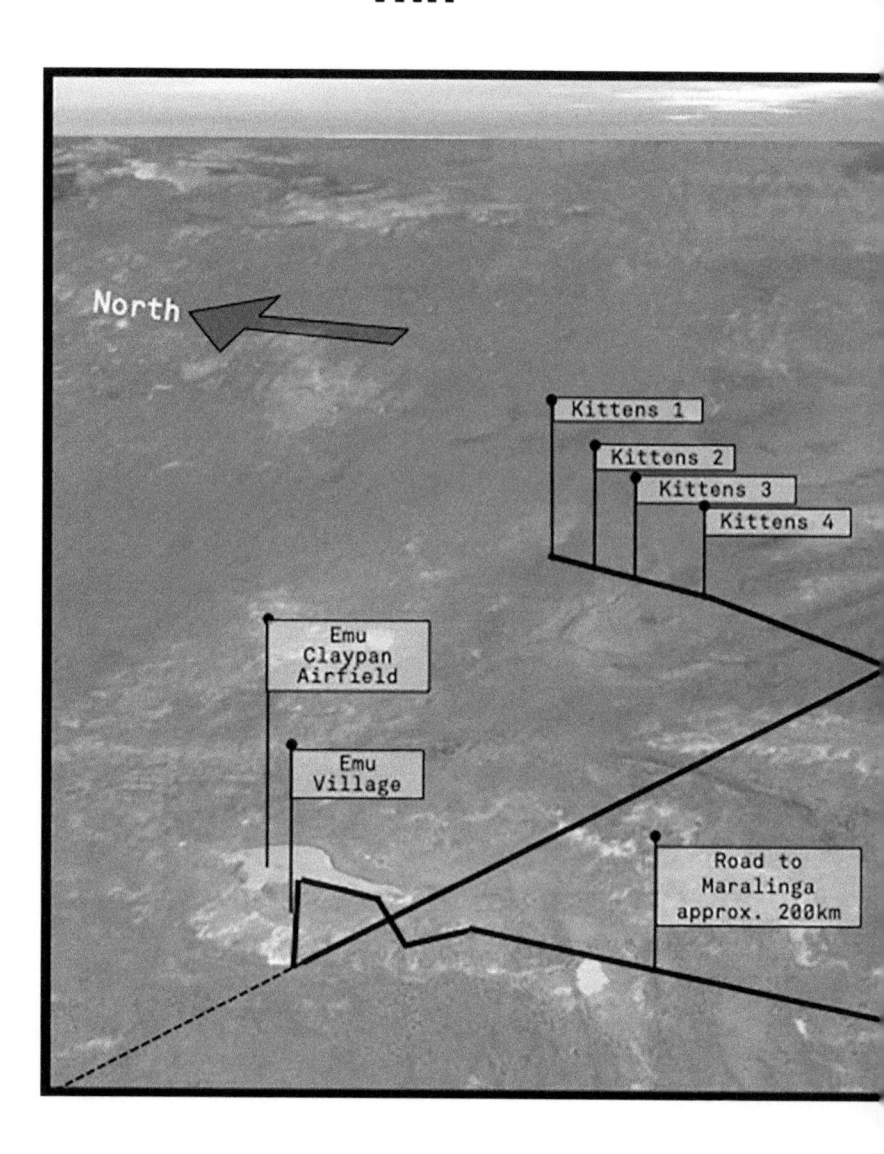

Emu Field Operation Totem 1953

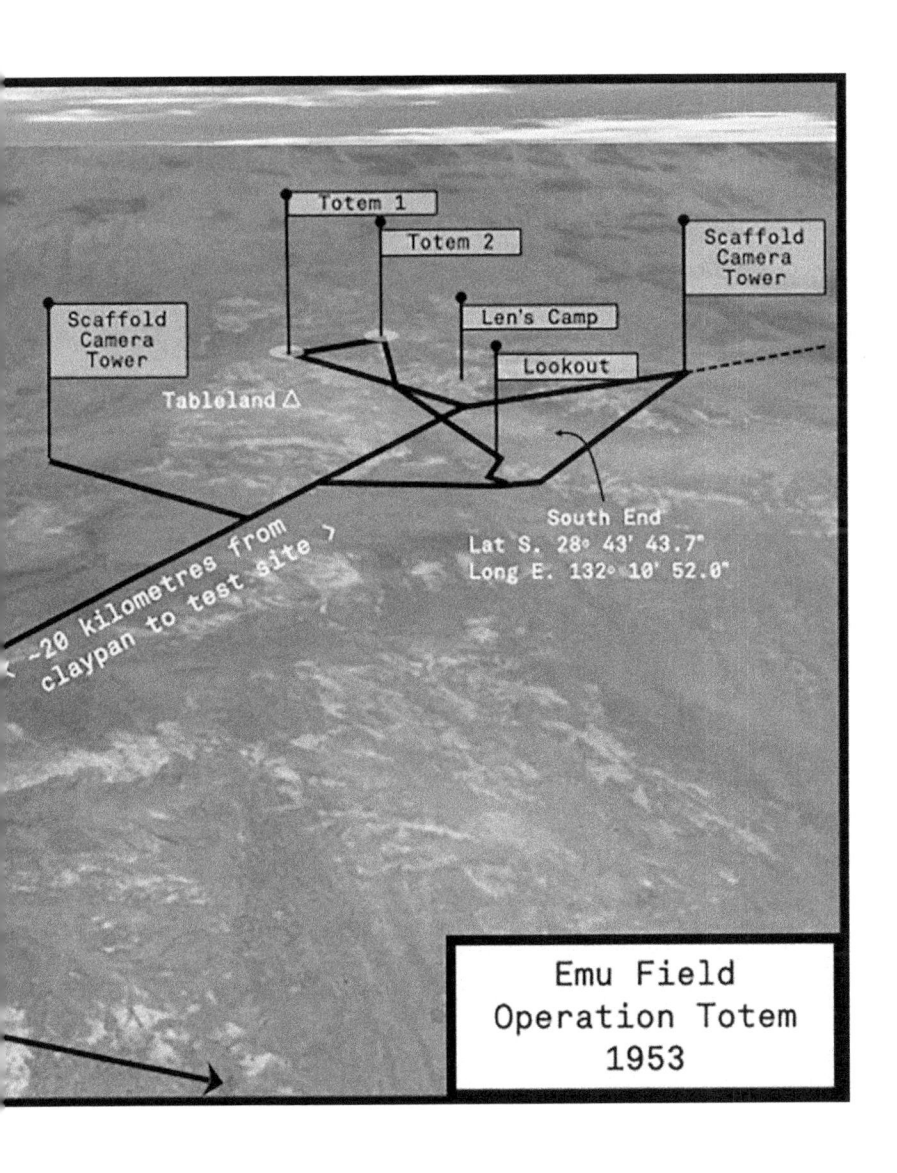

Totem 1
Totem 2
Scaffold Camera Tower
Scaffold Camera Tower
Len's Camp
Lookout
Tableland △
South End
Lat S. 28° 43' 43.7"
Long E. 132° 10' 52.0"

<- ~20 kilometres from claypan to test site ->

NB The exact location of the Kittens 5 test is not known.

SOURCE Dr Darren Holden

ABBREVIATIONS

AIRAC	Australian Ionising Radiation Advisory Committee
ANU	Australian National University
ARL	Australian Radiation Laboratory
ARPANSA	Australian Radiation Protection and Nuclear Safety Authority
AWRE	Atomic Weapons Research Establishment
AWTSC	Atomic Weapons Tests Safety Committee
CDA	Combined Development Agency
CRO	Commonwealth Relations Office
CSIR	Council for Scientific and Industrial Research
CSIRO	Commonwealth Scientific and Industrial Research Organisation
HER	High Explosives Research
LRWE	Long Range Weapons Establishment, Salisbury, SA
MARTAC	Maralinga Rehabilitation Technical Advisory Committee
NPO	Native patrol officer
PIPPA	Pressurised pile producing industrial power and plutonium
RAAF	Royal Australian Air Force
RAF	Royal Air Force
X200	Emu Field site surveyors' code

MEASUREMENTS

During the period of the Emu Field tests, Australia used imperial measurements:

1 inch = 2.54 centimetres
1 foot = 0.309 metres
1 yard = 0.914 metres
1 mile = 1.609 kilometres
1 ton = 0.907 tonnes

Note that because the international standard for altitude in aviation is feet rather than metres, feet has been used in this book to indicate altitude.

Until February 1966, Australian currency was pounds, shillings and pence. At the time of the changeover, one Australian pound equalled two Australian dollars and a shilling (12 pence) was the equivalent of 10 cents.

PROLOGUE

> From Arckaringa we'd ride back on the trucks to [Wintinna] and from there we'd walk to Wallatinna, south of Granite Downs. We'd follow the creek up the hills and walk across the bush, where our people knew the little places and were able to find water in the rockholes. However, before travelling out over the drier areas they'd wait for the thunder-storms to bring the rain to fill the rockholes, claypans and swamps.
>
> Yankunytjatjara man Yami Lester,
childhood survivor of Operation Totem.

The road between Emu Field and the much larger Maralinga atomic test site to the south runs for just under 200 kilometres, a rough track carved into two bumpy wheel lines. Every so often, drivers encounter drifts of fine reddish sand that cover the track and can cause vehicles to slip. The road is no longer open to the general public. When greater access was allowed, cultural sites along the way – notably the startling rise known as Observatory Hill – were mistreated and disrespected.

Vehicles average around 40 kilometres per hour on this road, so it takes at least five hours to drive between the two old atomic sites. The length of the drive is leavened by the sheer beauty of the surroundings. The black mulga trees dominate the landscape, taking on fantastical shapes, hewn by the winds, the availability of water and the bracing harshness of the environment. The land is mostly flat, and the sudden appearance of Observatory Hill, with its dramatic steep and rocky rise right next to the track, breaks up the plain and draws the eye.

The horizon seems impossibly far away, especially to people more used to the city, and on most days the sky is a vivid blue dome that sets off the salmon-red sand, both colours saturating the senses. The effect is beautiful and engaging. The shock and awe of the atomic tests have long faded, leaving a nearly silent and slowly healing landscape, still blighted by the remnants of the tests. At the halfway point between Maralinga and Emu, a particularly beautiful tree by the side of the road carries the mark of the man who carved out the road, the larrikin surveyor of central Australia, Len Beadell. He and his team built the road between Emu and Maralinga in 1955, after Emu had been abandoned and before the new Maralinga site began operations. Beadell was building roads all through the area at this time, and this particular road would keep options open for Emu either for future use of the site, or to retrieve infrastructure that had been left behind.

The shape of a gothic window has been hewn into the sturdy trunk of the tree and filled in with white paint, giving off a faint echo of a medieval church. The word EMU has been incised into the gothic window, and below sits an aluminium plaque, not the original (which was stolen years ago), that points back 60 miles to Maralinga and forward 60 miles to Emu. One can easily imagine Beadell and his men resting here during their surveying and road-building labours and feeling the need to mark their stay. After sharing billy tea and tinned corned beef, laughing all the while at Beadell's jokes, they would have packed up their things and moved on, taking readings and measurements, placing a Western grid over ancient land.

The land around Emu Field is both stark and enchanting, its long vistas and huge sky overwhelming, taking the new visitor by surprise. This small, hidden place is on the edge of the Great Victoria Desert, about 900 kilometres north-west from Adelaide. Its beauty does not make it any less harsh, a harshness that was the catalyst for the extraordinary skills of the people who had inhabited it for millennia, and who continue to visit and to care for this place. But no-one lingers here now. It's damaged.

Of course, this sliver of territory was not always called Emu Field. That label, bestowed by European outsiders in the 1950s, marks the tiny moment in time when Emu was the centre of the universe for a small group of atomic bomb makers and their political masters. Their story is complex and hidden, and its ramifications ongoing. Emu Field is marked on the map and in the minds of Aboriginal people as tainted. The two chunky concrete plinths at its atomic bomb ground zeroes are meant to outlast human memory. For whoever encounters this land in the far future, they signify the momentous events at this spot.

Today Emu and Maralinga are managed by Maralinga Tjarutja, the statutory body set up by the *Maralinga Tjarutja Land Rights Act 1984* to represent the Traditional Owners of both atomic sites. The traditional people of Maralinga and Anangu Pitjantjatjara Yankunytjatjara refer to themselves as Anangu in both Pitjantjatjara and Yankunytjatjara, two close dialects of the Western Desert. In October 1953, without any consultation with these people who had traversed these lands for millennia, the British government, with the full approval of its Australian counterpart, detonated two atomic devices at Emu Field under the codename Operation Totem. Material from those exploded bombs splayed outwards towards human settlements. The wrenching irony and tragedy of Operation Totem for both the Aboriginal people and the military personnel caught up in the tests were that the harm was caused in pursuit of technology that was soon to become obsolete.

Britain tested its atomic weapons in Australia between 1952 and 1963, with five major series that produced 12 'mushroom clouds'. The first was Operation Hurricane at Monte Bello Islands (also Montebello) off the coast of Western Australia in 1952. This was followed by the Totem series in 1953 at Emu Field. The British returned to Monte Bello for the Mosaic series, before moving to Maralinga for Buffalo in 1956 and Antler in 1957. In addition there were so-called 'minor trials' starting at Emu in 1953 and lasting until 1963 at Maralinga. The minor trials examined various aspects of atomic weapons technology but did not (for the most part) involve fission reactions. Following some ineffectual clean-ups at Maralinga during the 1960s, the British left behind a hidden toxic mess for the Australian Government to deal with many years later. No large-scale clean-ups took place at Emu.

Operation Totem, then, was the second British atomic test series in Australia. While Hurricane was well away from inhabited areas (although its fallout did reach people living on the mainland), Totem was right in the middle of Aboriginal homelands. The people most harmed by Totem lived in settlements and on stations that might be 150 kilometres or more away, yet they were still within reach of the noxious products of the tests.

Len Beadell drew the first Western maps of Emu. The area, hastily designated by the British bomb-makers as a test site, was entirely different to Hurricane's maritime site. Emu is indelibly terrestrial, although it lies close to the shores of a huge and ancient bay, long since turned to sand, the waters retreating millions of years ago to what is now known as the Great Australian Bight, well to the south.

The site was chosen because of its extraordinary, dramatic claypan. This orangey-brown stretch with its crazed clay is straight and long, and was used as an airstrip until Royal Australian Air Force (RAAF) labourers built a sturdier, all-weather runway right next to it. Water seeps through the cracks in the clay, though today there are no signs of the emu claw marks that gave the site its European name, or indeed of any emus.

Stretching off to the north-east and north-west, well beyond the current Maralinga Tjarutja lands, were small settlements, missions and pastoral stations populated mostly by Aboriginal people who trace their songlines back through thousands of years. That history has never quite been erased, despite the aggressive atomic colonialism that intruded into central Australia and left behind residues and fears for these people and their descendants. Many of the people affected by Totem were living well away from Emu Field, or were possibly caught unaware, as they traversed the land between waterholes and dwellings.

Today, the Oak Valley Rangers, an intrepid team of Aboriginal custodians, patrol the area and practise their cultural healing on this place, although the rangers themselves have no direct ties to the Emu land. This seems to accord with the view at the time of Britain's Operation Totem, at least in Walter MacDougall's understanding, that the claypan area, while in Yankunytjatjara country, held no special significance to the local Aboriginal people.[1] Of all the white people involved in the tests, Walter MacDougall, the Commonwealth Government's native patrol officer (NPO), was the most knowledgeable about Anangu lands. MacDougall made it clear that while Aboriginal people tended to stay close to the stations and other settlements for most of the year, during the dingo pup hunting season from August to October, they 'spread over the country'.[2] In early 1953, MacDougall had mapped 22 soaks or rock pools that Anangu used in the area during the hunting season, outside of the pastoral areas and therefore not controlled by white pastoralists, and conceded that there might have been more.[3] In October, when the tests were scheduled, Aboriginal hunters had access to water and were moving around.

Oak Valley is a small Aboriginal community, about a seven-hour drive from Emu. The fragile beginnings of the settlement in 1984 were the first sign that the people of the area were permanently returning after the dislocation of the atomic tests and other Western disruptions. The Royal Commission into British Nuclear Tests in Australia began in July that year, and the land had been returned to its Traditional Owners. Changes were afoot. The Oak Valley pioneers 'reflected the hard core aspirations of traditional Pitjantjatjara bushmen: 150 people living semi-traditional lives in one of the harshest homeland settings imaginable: no houses, few vehicles, no running water and no servicing facilities'.[4] The fledgling settlement took a while to solidify into the strongly traditional community it is today. From their base at Oak Valley, the Oak Valley Rangers now go forth to tend to their turf, undertaking many practical activities for Maralinga Tjarutja.

Ominously, there are almost no living creatures to be seen at Emu, except for swarms of sticky little flies, and the occasional feral camel; the dingoes are too canny to make themselves known to visitors, and who knows where the kangaroos and emus have gone. Almost no-one visits other than the rangers. Members of the public require a permit from both the Department of Defence (Emu is still part of the Woomera

Prohibited Area) and Maralinga Tjarutja to access the site, despite the fact that Emu is located on the grandly named Anne Beadell 'Highway', a 1325-kilometre stretch that connects the opal mining town of Coober Pedy in the east to Laverton in Western Australia. Named for Beadell's wife, the highway is a corrugated, two-wheel track that rattles the bones of anyone who travels on it.

Hundreds of Aboriginal people lived to the north-east and north-west of Emu in 1953, mostly on stations. Their lives were wrenched by uncaring outsiders who had no understanding of, or interest in, the richness of their culture. Many of them and their families have carried a heavy weight of anxiety since. Their ties to the land are damaged but enduring, and Anangu continue to pass down their stories.

One of the best-known storytellers was Yankunytjatjara man Yami Lester, who told of a place called Larry's Well, once known as Dolarinna. This ruggedly beautiful spot was not far from the settlement of Wallatinna, a station 484 kilometres due south of Alice Springs and over 170 kilometres north-east from Emu Field. Larry's Well had a creek with lots of young mulga trees, a favourite place before Totem for Aboriginal children like Yami:

> If you looked at it with the sun shining through the branches you could see the honey as liquid running down to the ground, right around the bottom of the tree. There it hardened and so we used to break it up into pieces to eat, or suck on it. We'd break it up, heap it on hard ground, cut the leaves off the branches and make a bundle of it for everyone to carry.[5]

In Lester's words, every square mile of this land was like a book.[6] The elders would read that book to the children to pass on the knowledge and the wisdom and the stories. Throughout his life, Yami Lester read that book to many children, and some of them continue as custodians and storytellers, although his voice went silent in 2017.

The central desert is A̲nangu country, where the most precious thing of all is water. The broader area today is known as the Great Victoria Desert, a colonial name honouring an imperialist monarch. But A̲nangu people have traced and mapped hidden water over thousands of years, and passed this vital knowledge on to succeeding generations. A̲nangu life has always revolved around water. Their artworks often show bold, unmistakable representations of waterholes and the well-trodden tracks between them. While the British bomb-makers endlessly fretted about the lack of water at Emu, A̲nangu could pinpoint it exactly. The British would not have thought to ask them, though, and A̲nangu people would rightly have been wary of sharing cultural knowledge with those who had come to ruin the land.

Central Australia may appear empty to anyone unaware of its true character, but it is filled with life and legend. The landscape is dominated by sand hills and distinctive desert vegetation. This glorious part of Australia is desert country, but it is not just sand; it is no Sahara. A̲nangu know the beautiful intricacies of the land in ways that are only possible after thousands of generations and 60 000 years. The culture, rich in storytelling and knowledge of natural resources, is predicated on movement. A̲nangu have been walking across the land throughout their long history, and are still moving in the modern era, 'hitching, in dodgy cars, in the Church bus, in Toyota troopies on bush trips – with an energy and frequency that is astounding', as scholar Eve Vincent puts it.[7] They know the land in ways that outsiders cannot. More than that, they know the sky above and the vast evening tapestry of stars, presided over by a bright moon.[8] They know how to find the hidden water and where the sacred places are. They understand its shadows and its light. They keep its stories and its songs, many in the form of impenetrable secrets that the uninitiated have no privileges to hear. None of that has changed, despite what happened there. Or, more accurately, the original inhabitants have changed the way they view the land of the Totem tests, but they continue to care for it all the same, nursing it back to health as best they can. The displacement first caused by the Emu Field tests, then worsened by what came later at Maralinga, has had a lasting, rippling effect throughout central Australia, down to the coastal communities of Yalata and Ceduna and to wherever Anangu have dispersed.

The atomic tests held at Emu were not the first incursion on Anangu people. The decline of traditional life on Anangu land occurred slowly at first, and largely without overt violence.[9] Instead, the tide of European encroachment, in the form of cattle and sheep properties, the churches and the railroad, all affected traditional patterns of life. One of the more benign was Ernabella, a Presbyterian medical mission established by Charles Duguid in the 1930s, to the north-west of what became Emu Field.[10] In the 1940s, its position was too close for comfort to the 'centre line' that governed the trajectory of the Woomera rocket tests. Duguid, a fierce and vocal critic of Woomera, was one of the few white people who took up the cause of Aboriginal displacement caused by the rocket testing, but his efforts had no effect. The British rockets flew through the huge blue sky, and later the atomic bombs sent both visible and invisible poisons in all directions.

During the decade after it was established, around 300 Anangu people came to live around the mission. Duguid, a

European doctor and missionary, saw Ernabella as a kind of buffer between traditional lifestyles and the relentless Westernisation that was changing the rhythms of life in the region.[11] Ernabella was perhaps unique among the local Christian missions in that it honoured and respected Anangu culture, and insisted that any white person who worked there learn its languages, primarily Pitjantjatjara and Yankunytjatjara.[12] In contrast, the United Aborigines Mission that operated at Ooldea, close to the site that would become Maralinga, insisted that the recipients of its largesse give up their language and culture.[13] Historian Dr Heather Goodall, a researcher attached to the legal team who represented the Aboriginal people at the Royal Commission in the mid-1980s, lived at Ernabella in the early 1980s. She observed that 'the Ernabella mission had a very strong commitment to support for and appreciation of Anangu culture'.[14] Duguid said in 1957, 'No one can deny that changes brought about by Government and commercial enterprises in the Central Aborigines Reserve in recent years have disturbed the balance with nature achieved by aborigines of Australia over thousands of years'.[15] While the atomic bombs were especially destructive, other forms of cultural destruction were entrenched by 1953.

Ernabella is the entry point into the atomic tests story of Walter MacDougall, a significant character associated with both Emu and Maralinga. He had tended sheep at the mission in the 1940s and did some lay preaching as well. MacDougall left the Ernabella Mission in 1947 to begin work as an NPO, employed by the Commonwealth Department of Supply, and based at the Weapons Research Establishment at Woomera. He was charged with communicating with people who lived and travelled over a 100 000-square-kilometre area, warning them of the rocket tests and later the atomic bomb blasts. Initially he travelled by train, although eventually he was given a car to bash through the bush. Between 1947 and 1956, he did this alone, until Robert Macaulay was appointed as a second NPO for the area.

MacDougall's story became partially entwined with that of the inhabitants of the central Australian lands. Well regarded by those who lived around Ernabella and the other Aboriginal settlements he visited, he was a kindly if austere and uncompromising man. He cared about the Aboriginal people he knew and lived alongside but, ultimately, could do little to save. MacDougall was known to many as Mr Mac and in evidence to the Royal Commission, Yami Lester described MacDougall as 'gorda', meaning 'brother'.[16] Still, MacDougall, one of the few white men who came to know Anangu people well, was not perhaps as much of an insider as he liked to believe. He was a trusted figure but not an initiate.

While the people living at Ernabella were not in the firing line of the fallout from the Emu tests, they nevertheless took on the anxiety that came from knowing that the lands were being destroyed in mysterious and frightening ways. Ernabella tells us about how the arrival of British weapons tests affected the psychological health of the Aboriginal inhabitants and how the sheer ignorance of the test authorities has been paid for in sorrow over generations. The stories of the tests intermingled with other catastrophic events, most notably the introduction of measles into these vulnerable communities in 1948, which, as Duguid told the Department of Aboriginal Affairs, was brought to the area by a child travelling on the Ghan train.[17] Another measles epidemic in 1957 at Ernabella added to the fear and uncertainty that pervaded the lands. These two measles epidemics wiped out up to one-third of Anangu living in and around Ernabella.[18]

Aside from MacDougall, the Europeans connected to the

atomic tests in Australia – both the British and the Australians – were ignorant of these Aboriginal homelands. Official information sent out to the media several months before the 1953 Operation Totem tests stated baldly, 'There are no aborigines in the area'.[19] In this way the white population of Australia was lulled, and few sought to ask any questions about the potential for the tests to harm Aboriginal people. Those who did speak up, notably Duguid and MacDougall, were largely ignored, or treated as cranks and sidelined.

To the north-east of Emu and east of Ernabella, the Granite Downs settlements that include Wallatinna (sometimes spelled Wallatina), Mintabie and Mabel Creek also have stories of community fracture caused by the Emu Field tests. Wallatinna is about 170 kilometres north-east of Emu, exactly the right distance, given the weather conditions, for a rolling cloud from Totem I to arrive around midday, about five hours after the 7 am blast. Meteorologists issued warnings not to fire the first of the Totem atomic bombs because of unfavourable conditions, and the bomb testers' own manual forbade testing in the conditions that prevailed. But the authorities fired anyway. In a split second, fates were sealed.

The name Operation Totem clearly relates to the idea of a totem pole. The British government's most fervent advocate of nuclear technology, Frederick Lindemann, also known as Lord Cherwell, bestowed the codename in December 1952, commenting, 'anything up the pole is appropriate'.[20] The Totem tests generated heat at the centre of over 6700 degrees Celsius and 6200 degrees Celsius respectively for the first and second blasts, hotter than the surface of the sun.[21] The explosions themselves mimicked the brightness of our life-giving star and were heard hundreds of kilometres away.[22] As RAAF historian Ivan Southwell wrote in 1962:

At 300 miles [from Woomera], in absolute desolation, scorched and fused by blinding heat, still slightly radioactive, is Emu Field. There one steps warily or doesn't step at all. There, sprayed with yellow paint and silent in the sand, are abandoned trucks and jeeps and weapons once too hot to handle. There, near the bomb towers that vanished, the very surface of the desert has become as glass.[23]

The bomb towers made of steel atomised in the blast. Their substance was sprayed around, possibly joining the radioactive products of the explosions to create a truly toxic mixture. Since the Totem tests, both ground zeroes have featured trinitite, a form of glass created by superheating the soil and limestone, still mildly radioactive.[24] Trinitite is named after the 1945 Trinity test at Alamogordo, New Mexico, culmination of the Manhattan Project. The glass of Emu Field is one of its most enduring and damning features. It's still there on the ground, nearly black but glinting in the sun.

In the mid-1980s, decades after the tests at Emu, the Royal Commission into British Nuclear Tests in Australia held hearings in central Australia. The Royal Commission was established by a reluctant Peter Walsh, then minister for resources and energy in the Hawke Labor government. Walsh resisted at first, convinced it would be an expensive waste of time that would only benefit the lawyers.[25] However, by May 1984 so much disturbing evidence had come to light about the contamination left behind by the tests, including a scientific survey by the Australian Radiation Laboratory (ARL) and allegations aired publicly by Yami Lester, that his hand was forced. The commission was announced in July 1984 and began taking evidence in August. The report was delivered to the federal government in November 1985. It was headed by the flamboyant James McClelland, often known as Diamond Jim, who had been a minister in the Whitlam Labor government in the early 1970s. At the time of the Royal Commission, McClelland was chief judge in the Land and Environment Court of New South Wales. His fellow commissioners were Jill Fitch, senior health physicist for the South Australian Health Commission, and William Jonas, a proud Worimi man, then lecturer in geography at the University of Newcastle and later the Aboriginal and Torres Strait Islander social justice commissioner and race discrimination commissioner. This was an especially peripatetic judicial enquiry, with oral evidence taken in Sydney, Brisbane, Melbourne, Adelaide, London and Perth, plus the remote locations Marla Bore, Wallatinna and Maralinga in South Australia, and Karratha in Western Australia.

A dozen or so Aboriginal people who had been present at Wallatinna during the Totem tests gathered in the desert sand with Royal Commission president Jim McClelland to tell their stories, many of them through interpreters. Some evidence, including Yami Lester's, was taken at Marla Bore and some at Wallatinna, 20 kilometres away. Their stories began with accounts of how the men would catch bush meat and the women would gather quandongs and wild tomatoes, to supplement the food they got from the station.[26] The stories culminated in harrowing accounts of a greasy black miasma that blanketed their communities and left a legacy of illness and unease.

The Royal Commission came to the desert in the 1980s because of a complex and intertwined set of circumstances. A growing land rights movement had shaken the complacency of the white population of Australia, introducing the possibility of Aboriginal people claiming back some of what had been taken from them by British colonisation. Gough Whitlam, Australia's crash-or-crash-through prime minister, introduced land rights legislation into federal parliament during his tumultuous tenure in the early 1970s. When the constitutional crisis of 1975 removed him from office before the legislation had passed, it was picked up by Malcolm Fraser, the prime minister who ousted him. Over time land rights became the law of the land.

In 1984, Maralinga Tjarutja received freehold title to 76 000 square kilometres of land in outback South Australia, encompassing both the Maralinga and Emu test sites.[27] Adjacent lands were also handed back to Aboriginal owners. The return of the lands was not without a fight. Anangu developed a political consciousness as land rights activism rose throughout Australia. A commemoration ceremony held in October 1983 at Wallatinna, exactly 30 years after Totem I, was a rallying point.[28] Yami Lester took a video recording of the event with him to London in 1984, and the British government got its first inkling of the anger and accusation that were coming its way.

At the time of the tests, the British had affected not to know about Aboriginal people and were quite content to leave any complications relating to them to the Australian Government. The Royal Commission followed Yami Lester to London in early 1985 and took evidence from many of the surviving British participants, most notably the head of the test series, Lord Penney (Professor William Penney), by then elderly and unwell, and far removed in every way from the events in the Australian desert. Penney, credited as the father of the British atomic bomb, said in evidence to the Royal Commission in London, 'The line I took was that overall, the safety of the range must be a British responsibility but we knew nothing about Aboriginals and would the Australians assume the responsibility of being sure that there were no Aboriginals at the time of firing. I think that was agreed'.[29] That convenient line of reasoning only went so far. The Royal Commission did demand and get some answers. Insufficient though they were, given the enormity of the events at Emu Field, they began a process of delving, processing and understanding that continues, gradually bringing the secrets of Emu Field into the open.

Chapter 1
FINDING X200

> An important purpose is at last being served by Australia's notorious Dead Heart, although it would be impious to suppose that our central desert was designed for a purpose in some senses so sanguinary. It seems probable that, except in the polar regions, where the climatic conditions would add tremendously to the difficulties to be surmounted, there is no part of the earth's surface that would lend itself to long distance atomic tests so well as the arid wilderness that extends from northernmost South Australia across the territory of her western neighbor, and does not end, considered as a rocket range, even when it reaches the Indian Ocean.
>
> 'A British Nevada', Adelaide *Advertiser*.

In early October 1952, the British government under Conservative prime minister Winston Churchill was well pleased. The country had just demonstrated its atomic weapons capability by detonating a device of its own design at Monte Bello, a remote island group off the coast of Western Australia. Operation Hurricane had provided a much-needed fillip to a country still under the pall of a devastating war; more importantly, its major ally, the United States, began to show some grudging respect that the United Kingdom had built its own A-bomb. The weapon, known as Blue Danube, was a plutonium-based fission weapon modelled on the bomb dropped on Nagasaki in 1945, but with features that made it unique to the UK atomic weapons program.

Nuclear weapons technology emerged dramatically from the clandestine Manhattan Project that created the atomic bombs dropped on Hiroshima and Nagasaki in August 1945. The United States tried to monopolise this technology through strict, paranoid legislation known as the McMahon Act (officially the Atomic Energy Act 1946), which banned America from collaborating on atomic weapons with any other country. The intention was to stop the Soviet Union from developing nuclear technology, and it failed spectacularly, largely thanks to the efforts of spies among the Manhattan Project scientists. But since the Act excluded allies as well as the enemy of the day, Britain was left by the wayside as the United States forged ahead, rapidly turning the technology developed during the Manhattan Project into an extensive, deadly arsenal. After some deliberation, Britain decided it was having none of this and instead would bootstrap the knowledge of its scientists and build its own atomic weapons. To do so, it would need to exploit its former colonies, and one in particular answered the call.

Britain managed without difficulty to secure the agreement of Robert Menzies, Australia's prime minister, who in 1950 granted access to Australian territory for the first atomic test without initially seeking approval from his own Cabinet. Indeed, only two of his ministers knew about the plans until just before the test occurred. Once Operation Hurricane was chalked up as a success, Britain was not about to stop. With the patriotic boost from Hurricane, plans soon crystallised for the next stage in the development of nuclear capability. Australia was again in the sights of the decision-makers in London, though the country was far from their first choice. Despite casting a large net to find places to test its innovative A-bomb designs, Britain failed to produce many suitable catches.

From the early days of atomic weaponry, finding testing sites has been fraught. Few places on Earth are suitable, taking into account factors such as human habitation, geographical and geological features, access to ports and infrastructure, meteorological conditions and water supply. Indeed, most of these considerations were not satisfied by the flat claypan and associated scrub land that came to be known as Emu Field. This rugged, remote site was often referred to as X200 in secret documents about the tests, following the surveyors' initial designation (Maralinga was X300). Emu was found and used and then discarded, hurriedly. The only lasting physical infrastructure is a brutally rough track running 232 kilometres eastwards joining Emu to Mabel Creek Homestead, and another 193 kilometres south to Maralinga.[1] Australia spent tens of millions of dollars in today's money to establish a site that was only used for a brief, frenetic moment in 1953. Expensive though it was, the human cost was considerably higher and the damage long lasting.

The story of Emu Field is of enormous significance to Australia, and the British authorities have gone out of their way to bar access to its central secrets. As Justice James McClelland wrote in his Royal Commission report, the Emu tests were from the beginning 'shrouded in a veil of secrecy which, by virtue of the lack of contemporary documentation is hard to penetrate'.[2] The British continue this strict control of all information about Totem. At the time of writing, the main UK files pertaining to British atomic testing in Australia, the ES and AB series at the

National Archives, are no longer available for public viewing after a mass withdrawal of files in December 2018. Even before the withdrawal, British files about Emu were particularly hard to get, with whole files or parts of files heavily redacted, officially described as 'retained by the Ministry of Defence'; some files were redacted after being publicly available for many years. The full story in all its technical details may never be publicly known.

For example, ES 1/834, a file that deals with contemporary Operation Totem reports, was partially available for viewing in early 2018, but completely withdrawn by the end of that year. It includes a report by E Drake Seager, group leader, Army Equipment Group of the UK War Office for Operation Totem (and later commander of the Indoctrinee Force at Maralinga, a group of officers deliberately placed close to weapons tests in 1956). Drake Seager's report sheds some light on Totem's extraordinary secrecy, and the system designed to make it harder to glean specifics:

> Reports should not reveal information from which the powers of the explosions could be deduced unless it is part of the group responsibility to obtain such information. In general this objective can be achieved if one parameter only is quoted (e.g. heat radiation but not heat radiation and distance, blast pressure but not pressure and duration or pressure and distance). Contamination levels at points defined topographically, or by distance references, should not be quoted unless essential in describing results which the group has been authorised to obtain.[3]

Drake Seager advised that any report concerning the results of an atomic weapons trial should receive the classification

GUARD, when even the existence of the document could not be disclosed to anyone without the highest security clearance. The classification SECRET – GUARD was to be used for any reports giving one phenomenological parameter (that is, either heat radiance or distance). The classification TOP SECRET – GUARD was to be given to any report containing both parameters.[4] The British did for a time allow files with those evocative words stamped all over them to be released; more recently, they have drawn them back into the shadows. Yes, all contemporary atomic weapons development information was secret, and this is reasonable during weapons development. But why the continuing secrecy, all these years later? The vacuum in the documentary record around Totem represents the very worst impulses of the British government, to control, deny and cover up its activities, even retrospectively and in relation to Australian territory, where Australians were affected.

Britain did not trust Australia at all when it came to atomic secrets and had a spy in Australia, presumably an MI6 agent. A top-secret February 1953 meeting within the UK Ministry of Supply discussed security aspects of the forthcoming Totem trials:

> Sir Thomas Elmhirst has ascertained that our own security service has an agent in Australia; Mr [Neil] Pritchard [assistant under-secretary for Commonwealth relations, later UK high commissioner to Australia] is writing to our Deputy High Commissioner in Australia to arrange for that agent to be put into the 'TOTEM' picture, and instructed to report to us if he has any reason to suppose that local arrangements are not satisfactory.[5]

Air Marshal Thomas Elmhirst was the head of TOTEX, the UK committee that oversaw planning for Totem.

Quite a bit was known about the Emu site, even before Operation Hurricane in October 1952, when Britain became the third country to test an atomic weapon (after the United States and the Soviet Union). The UK Ministry of Supply personnel had taken a keen interest in a mainland site along the centre line of the Woomera rocket testing range as early as June 1952, when Len Beadell led a reconnaissance operation.[6] In September, Dr William Penney, then chief superintendent of High Explosives Research (HER, from 1953 the Atomic Weapons Research Establishment, or AWRE), visited the X200 site and pronounced it difficult but suitable. He estimated that, because of its physical limitations, the total number of personnel at Emu would not exceed 300.[7]

A few months earlier, a top-secret conversation between Penney, Alan Butement, the chief scientist for the Australian Department of Supply, and Professor Leslie Martin, the Australian defence scientist, may have sealed the fate of the Aboriginal populations in the vicinity of Emu Field. At the time, the wider Australian Government had only recently been made aware of Operation Hurricane, while Britain was still rather desperately negotiating behind the scenes to test at American sites. However, the uncovering of the nuclear spy Klaus Fuchs in 1950 had spooked the Americans and made co-operation politically difficult. So in April 1952, Penney confided in two security-cleared Australians that although Hurricane was 'special' and Monte Bello matched its particular characteristics (Hurricane was deliberately designed as a maritime test of a bomb exploded in the hull of a ship), that was not the end of the matter. Britain was in for the long haul, and '[f]urther tests would be simpler in conception and less costly in scientific man-power and general logistics'.[8] Penney wanted a terrestrial test site, and he wanted it quickly.

Penney was a British veteran of the wartime Manhattan Project. He would have preferred to do without the rigmarole of creating new atomic test sites when the Americans already had excellent sites that he knew. However, he had no choice. There was never any chance of the British homeland itself being sullied. 'The tests could not be carried out in the British Isles, but whether the U.K. would request the use of a test site from the U.S.A. or Canada, or Australia or elsewhere would, of course, depend on high level policy', according to the notes of the secret conversation.[9] After Penney laid out his requirements for remoteness, aircraft capability and laboratory facilities, Butement and Martin immediately suggested the Woomera rocket range, where the British had been testing rocket technology since 1947, and infrastructure was already in place. They noted some currently unoccupied buildings at Salisbury to the north of Adelaide, the home of the Australian Long Range Weapons Establishment (LRWE, later to become the Weapons Research Establishment), and adjacent to where the new Edinburgh RAAF base was about to be built. The notes of the Penney–Butement–Martin meeting state, '[I]n view of the importance which the Australian Government would undoubtedly attach to whole-hearted collaboration with the U.K. on testing atomic weapons, Professor Martin undertook to recommend that these buildings at Salisbury were held in reserve until the U.K. had clarified its policy'.[10] The existence of Woomera made the decision easier for Penney.

By mid-1952, a remote part of the north-western Woomera range had come into focus. Because the Woomera Prohibited Area offered existing facilities, the proposed test series, Operation Totem (until December 1952 codenamed

Tom-Cat), could be staged quickly. Butement and Beadell pinpointed the exact site from a vast and seemingly featureless swathe of land. On 18 September 1952, before Hurricane, William Penney visited the site they found, accompanied by the Canadian defence scientist Omond Solandt, along with Butement and several others.[11] Solandt would be an observer of Hurricane, and soon after proclaimed in the media that 'Britain now was a factor to be reckoned with in the field of atomic warfare'.[12] The group took on some heavy-duty bush-bashing to reach the site. They looked further afield at some alternative sites, and also ventured to Oodnadatta (320 kilometres away) to catch a plane back to Woomera. The terrain was heavily wooded and filled with sand ridges. This was no easy journey.[13]

From Woomera Penney went on to Western Australia. Once he had boarded HMS *Campania*, the vessel from which he initiated and witnessed Operation Hurricane, he cabled the Tory politician and scientist Frederick Lindemann. Lindemann was Churchill's atomic bulldog, and Penney wanted to let him know that he favoured going ahead with an atomic test site at a site adjacent to the newly-discovered claypan. That cable set in train a period of frenetic activity. In the same cable, Penney also envisaged more atomic tests beyond Emu, even before he knew his Blue Danube atomic weapon would work.[14] The successful test at Monte Bello was just the beginning.

On 13 December 1952, the formalities for establishing Operation Totem were observed in an *aide memoire* Lindemann handed to Prime Minister Menzies in London. The brief note said in part:

> Her Majesty's Government in the United Kingdom
> seeks the agreement and co-operation of Her Majesty's
> Government in Australia for the carrying out of an Atomic

Bomb test in October 1953 at Emu Field, 300 miles northwest of Woomera. The test has become necessary in order not to delay the rapid development of the United Kingdom Atomic Energy Programme. It is intended to explode two bombs, each considerably less powerful than that detonated at Monte Bello.[15]

The *aide memoire* was transmitted to the acting prime minister, Arthur Fadden, who asked the Minister for Defence, Philip McBride, to reply on 15 December 1952 with a simple message: 'Acting Prime Minister has authorised me to say that he has conferred with the minister for Defence and that they confirm that the Australian Government will agree in principle to the United Kingdom proposal'.[16] The amount of information conveyed to the Australians was minimal. Nevertheless, approval was granted. The die was cast.

Now the practical matters had to be dealt with, with little time to spare. Len Beadell had been the natural choice to survey a possible test site, since he had already surveyed the vast Woomera range. Described by researcher Eve Vincent as 'a kind of nuclear-age nomad, methodically recording longitude and latitude readings and making topographical notes', Beadell was ready and willing to do this work.[17] Once Emu was formally selected, he had a continuing role in setting up the site. Beadell relished being trusted with this secret task, and made much of it later when he wrote books and toured the country giving witty speeches about his experiences, slipping off the bonds of secrecy with much enthusiasm when he was finally able to do so. Some say that he gilded the lily somewhat in the claims he made about his role at the Emu and Maralinga sites, but the documentary evidence supports the fact that he was a major player in situating the mainland British atomic tests.

The British were beguiled by Beadell. After leaving Emu once Totem was complete, Penney wrote to the head of the Australian Totem Panel, Major-General JES (Jack) Stevens: 'Mr. BEADELL gave the greatest pleasure, perhaps, of all Australians present. His surveying skill made a deep impression on us, while his combination of bushmanship, artistic abilities and sense of humour did much to make our stay at Emu memorable'.[18] As RAAF historian Ivan Southwell commented in his book on Woomera, 'Beadell enjoys his fame and has never been known to make the least of a story'.[19] By all accounts, he was an amusing as well as a competent person, although, distressingly, inclined to ignore that his activities took place in the homeland of a colonised and marginalised people who would bear the brunt of the damage done. He seemed to suffer no discomfort about that.

In evidence to the Royal Commission, Beadell said that he had been told to look for a site about 482 kilometres northwest of Woomera, and not closer than 32 kilometres from the 'centre line'. He explained his work to establish the rocket range: 'In doing Woomera I had to arrange a centre line of fire [on] which we could arrange all the instrumentation that we were going to track the rockets and that was a certain line across Australia which I calculated and everything I did revolved around that line of fire'.[20] The centre line had to be kept clear because atomic bomb testing could interfere with the tracking of British rocket tests. He gave evidence that he searched just over 32 000 square kilometres on his own (although his colleague Bill Lloyd travelled with him in a separate vehicle), trying to find a site free of sand hills. Some of the sand ridges in that part of Australia extended for over 150 kilometres without a break. 'I did not know what a bomb site was supposed to look like but I knew that we had to have room to move without sand hills in front of you everywhere', he told the Royal Commission.[21]

Beadell and Lloyd travelled west from Coober Pedy, going past Mabel Creek and on to Tallaringa Well.[22] They took a right turn at the well and drove on for 100 kilometres before reaching a claypan called Dingo. According to a chronicler of Beadell's exploits, Ian Bayly, this claypan, while not suitable as a landing strip for aircraft, became a base for the search. Making use of Australia's sole available helicopter, Beadell cast his eyes over the terrain around Dingo, searching for a claypan that could become a landing strip and other features such as a site for a small village and a quarry.[23] Eventually Beadell and Lloyd found a 1.6-kilometre-long sheet of clay – 'a perfect natural runway in the middle of this wilderness' – dotted with emu claw marks.[24] Beadell camped at the claypan and took many kinds of measurements, including 'astro fixes' at night, plotting the exact location so that he would not later have to duplicate the effort of finding it. The huge chunk of country looked so much the same that this was a real risk, he told the commission, so he had to precisely establish the location.[25]

Department of Supply chief scientist Alan Butement knew the Dingo claypan from the time the centre line was first surveyed by Beadell in 1947, and he mentioned this to William Penney during a meeting in London before Operation Hurricane.[26] While Penney had decided the Dingo claypan was not suitable, he was interested in the newly discovered claypan, to the south-west. Soon the emu claw marks would provide its name. The site is on a plain, surrounded by low mesas. The terrain is sparsely vegetated by small trees, mostly mulga, and is dotted here and there by the epic sand dunes of the Great Victoria Desert.[27] The soil is sandy, with plenty of limestone poking out.

Various aspects of the site were examined ahead of the atomic tests, such as 'the texture of sand and ground, and, particularly ... detailed data of meteorological conditions', according to a letter from Ben Cockram at the UK High Commission to Australia to Neil Pritchard in the Commonwealth and Foreign Office.[28] These observations found that '[it] was likely that meteorological conditions would also be proved to be more tricky than at Monte Bello as regards the possibility of contamination of other areas as a result of a weapon test'.[29] A handwritten exclamation in the margin of this letter, 'No!', to the idea expressed that Emu might be used for more tests after 1953, seems to confirm that this was never the plan and that Operation Totem was only ever to be used for these very particular, unique tests.[30]

The extension required to the Woomera Prohibited Area to accommodate Emu Field was kept deliberately low key and quietly slipped through the official processes in March 1953. The head of the Prime Minister's Department, Allen Brown, wrote to Ben Cockram to let him know that the process had gone smoothly and silently: 'You will be interested to know that the additional prohibited area at Woomera has been proclaimed. The Proclamation was published in the Commonwealth Gazette on the 12th March. As you will see, it has effected no comment'.[31] The less comment the better. The UK authorities did not want publicity.

Cockram had written to Prime Minister Menzies on 3 March 1953, clarifying aspects of the operation and asking some questions. The letter was summarised in a briefing document in the Prime Minister's Department. Cockram had advised that a cover story would be needed for anyone snooping around, wondering what was going on in the far reaches of the Woomera range, to have them think that it was simply 'normal rocket range work'.[32] However, 'if and when speculation arises as to the high level of activity, there would be an arranged leak about an expendable pilotless bomber'.[33] The briefing document noted it was not yet clear whether the United Kingdom or Australia would arrange the leak. Cockram went on to say that 'the United Kingdom authorities will rely on Australia for a warning as to the progressive needs for cover plans. The Australian security people are alert to this need and will keep the Australian Totem panel continuously informed'.[34] In his letter Cockram had emphasised the point that British authorities wanted to defer any announcement about the tests at Emu until the latest possible date, 'so as to leave the minimum amount of time for use to be made of this information by outside interests, friendly or otherwise'.[35]

Acting Prime Minister Arthur Fadden told his fellow ministers about the forthcoming Totem tests in a memorandum drafted on 22 June 1953, less than four months before Totem I. He told them, 'I am sending you this note because the story is likely to break in the press at an early date and it may become necessary to forestall the press stories by an official announcement which will be issued simultaneously in London and Canberra'.[36] As always, Australia followed the secretive lead of the United Kingdom and tried to control public information about the tests. Indeed, the Australian Government tried to curry favour with the UK government by showing just how secretive it could be.

Emu Field had so little apparent water, it is a wonder it was considered viable at all. An Australian Department of Supply document outlined some of the background, saying '[t]he site was a very isolated one, and the time available for the preparation of the site was definitely limited. Water was a determining factor as to the size of the force working at Emu

Field, and conditions under which work was carried out were primitive and exacting'.[37] He assured his colleagues that 'the Australian Government is satisfying itself as to the complete safety of the operation'.[38] The Australian Government turned out to be easily satisfied.

Australian involvement in Operation Totem was deliberately kept to a minimum by the British authorities, although they sought to establish a veneer of Australian involvement, as a sop to the nation's pride. In a secret letter to Allen Brown, the physicist Ernest Titterton set out the ways that Australia could be involved, listing three categories of experiment, of increasing technical and security importance. Titterton, only recently arrived from Britain to take up the foundation chair of nuclear physics at the new Australian National University (ANU) in Canberra, was destined to become infamous as Australia's 'Dr Strangelove' – McClelland described him that way in his autobiography.[39] He told Brown, 'If Australia is to seek participation in the tests I believe it would be wise, initially at any rate, to confine our efforts to certain small projects which we could do effectively and well'.[40] Naturally, Titterton placed himself at the heart of those projects, and his experiments at Emu were to be the only 'Australian' research carried out at the site. Interestingly, Titterton was often referred to as an Australian physicist, but he had only been in Australia since late 1950 and was loyal to Britain. His 'Australianness' seems to have been conferred entirely by his academic position at the ANU.

Titterton's categories were intended to create a dividing line between insiders and outsiders. Category A included meteorological work, which had little in the way of security implications. Australia was welcome to operate in this category, said Titterton, and could engage two meteorologists, backed up by Commonwealth resources, just as they had done at Monte Bello. In the same category, Titterton suggested that Australians could be involved in radiological survey and monitoring work. He suggested a team of six people for the task, and offered training facilities at his ANU department if necessary.

Category B was a higher technical and security level, which encompassed his own proposed Operation Totem project – measuring the fast neutron flux as a function of distance from the weapon being tested, experiments that would help to establish the explosive power of the weapon. Titterton would lead a team of specially selected and trained junior scientists and technicians.[41]

Category C involved full weapons knowledge and both British and American security clearance. Here Titterton placed himself into a class above: 'It is doubtful whether the U.K. would admit anyone to this work who had not already been in it. I believe I am the only individual in the Commonwealth with the necessary clearance and experience'.[42] He went on to say that William Penney had asked him to be technical director for Operation Hurricane at Monte Bello, but he had declined because of his ANU commitments.[43] 'However, if it were considered important enough from the Australian point of view and was not too expensive in time I would be prepared to work closely with one of Penney's teams in the field or, if necessary, take charge of the test programme site in Australia'.[44] Titterton was in his element, enjoying his status as a member of the UK nuclear elite. He had, after all, been part of the British Mission to the Manhattan Project, an insider. When he visited the United Kingdom in 1953 for discussions about the forthcoming tests at Emu, he was invited to stay with William Penney and his wife. Penney had sent him a note on his distinctive blue notepaper, saying 'Joan and I would be delighted if you could stay with us while we are discussing technical matters.'[45]

There was another motive for involving Titterton in Totem. He was the only 'Australian' allowed anywhere near the core sensitive data about the test, although defence scientist Leslie Martin was given some privileged information. But Titterton, a sound British physicist trusted by his fellow countrymen, was going to oversee a scientific operation at Totem with the help of a team of three young physicists who might one day form the 'Australian nucleus if our function or interest in atomic functions or weapons ever broadened', a point recorded in a file note made by assistant secretary in the Prime Minister's Department, Allan McKnight, after Titterton had come to see him in April 1953.[46] Titterton seemed to be putting together a team of scientists who could lead a future Australian atomic bomb building effort.

McKnight's file note is interesting in many ways, not least because it reveals Leslie Martin's lack of enthusiasm about the growing British test program in Australia. 'He told me again of his interview with Martin in Melbourne some 4 to 6 weeks ago when he had found that Martin (who had previously discussed it with Butement) was very luke warm about the whole business of the test and Australian technical participation in particular.'[47] Martin was not keen on physicists from his lab at the University of Melbourne being involved with Totem. Titterton had tried to convince him otherwise, but in the end made use of the services of young ANU physicists. Martin had been described earlier by Penney as 'able, sensible and discreet', although he was never completely taken into the confidence of the British atomic scientists, as was Titterton.[48] The Australian minister for Supply, Howard Beale, who oversaw the Australian end of the British tests, noted in a top-secret

Cabinet document in 1954 that 'the degree of [Australian] scientific participation [at Emu] was greater than at Monte Bello, but nevertheless was limited to a small number of Australian scientists'.[49]

According to McKnight, Titterton made the case that his proposed experiments at Totem should constitute a substantive Australian scientific activity 'so that we would be something other than labourers and yardmen in the operation'.[50] It would then, Titterton suggested, be a matter for Australia to choose who among its qualified workforce should be involved. The undertone suggests that Titterton was the obvious person to lead such research, and there appears no real doubt that he would be the 'Australian' most involved. Operation Totem was a positioning opportunity for Titterton, a noted booster of both atomic weaponry and energy. This was Titterton's moment to hoist his flag on the territory of Australia's atomic future, and he did so with gusto.

On 11 February 1953, an advance party of one RAAF officer and 17 men arrived at X200 by air and started the mammoth process of building a test site from scratch.[51] The desert was dominated by mulga scrub and sandy soil, and not much else. The first construction personnel arrived on 14 March 1953, preparing to build 40 kilometres of roads, campsites and technical areas. By the end of March, 165 officers and men were engaged in construction work.[52] They lived in canvas tents with little water and primitive sanitary arrangements, constantly attacked by insects and challenged by the heat. The party initially worked 10 hours per day, seven days per week, until mid-May, when workers were granted half a day off on Sundays.[53]

The entire Woomera range was already proclaimed a prohibited area, but the government had to decide whether this was sufficient legal protection given the secrecy of the Totem tests. Advice was sought on whether to go a step further and invoke the new *Defence (Special Undertakings) Act 1952*. This Act, rushed through for Operation Hurricane, gave the Commonwealth the right to exclude everyone except security-cleared individuals from designated Australian territory. According to advice provided in April 1953 to the Totem Panel by the assistant crown solicitor, LG Egan:

> Under the Defence (Special Undertakings) Act it is an offence, not only to be in or to enter the Prohibited Area without a permit, but it is also an offence to fly over the Area unless a permit has been issued. The Act also makes it an offence to make a photograph, sketch, plan, model, article, note or other document of or relating to a Prohibited Area or anything in a Prohibited Area (Section 9), or to be in possession of a camera without authority in a Prohibited Area (Section 17), and also photographs etc. made in contravention of the Act are forfeited to the Commonwealth (Section 25). The penalties for offences against the Act range from imprisonment for 2 years to imprisonment for 7 years.[54]

Egan advised that 'the administration of a project at Woomera would be strengthened if recourse were had to the provisions of the Defence (Special Undertakings) Act'.[55] He said that the activities at Emu would have to be declared a Special Defence Undertaking by the minister for defence, and an officer-in-charge appointed (a role that could be carried out by the site superintendent). The sweeping powers of the Act would help guarantee and reinforce the secrecy of what went on there.

The site had earlier been declared a prohibited area under Regulation 90 of the *Supply and Development Act 1939–1948*. A top-secret memorandum sent to the Australian Totem Panel by public servant Leo Carter in March 1953 gave information about the ramifications of declaring it a prohibited area. 'Regulation 90 makes it an offence to enter, be or remain in a prohibited area except with permission. The regulation provides for the removal of unauthorized persons from the area and provides a small penalty of £50, imprisonment for three months, or both.'[56] Carter advised that the regulations under the Supply and Development Act were not strong enough, so the Defence (Special Undertakings) Act needed to be invoked:

> The *Defence (Special Undertakings) Act 1952* was, as you were aware, specially designed to cater for an operation similar to 'Totem', and experience in its working in relation to the atomic weapon test at Monte Bello Island demonstrated both its value and the necessity for the kind of protection which it provides in such circumstances. This Act gives a wide range of powers and authorities to enable an area prohibited under it to be protected both from the idle trespasser and the potential enemy agent.[57]

The Emu site was hermetically sealed by these new legal powers. No-one could enter without approval, although in reality the site was hardly impermeable. The possibility remains that Aboriginal people did enter, oblivious to any legal act or requirement.

Penney issued strict instructions to all UK staff before Operation Totem that they were not to have loose lips:

> Everyone concerned in the present atomic weapon tests will be in possession of some information which must be kept to the person concerned and which would cause damage if it gets into the wrong hands. The following notes are intended for your guidance on what you may and may not say. If you are uncertain about any point then SAY NOTHING.[58]

He continued:

> You may quote in conversation or in letters home or to your friends anything obtained from the release given to the Press and, in general terms, anything that you saw. Do not be precise about distances even if you know them as they may convey valuable information in this way ... Do got give the distance at which certain Service or Civil Defence Equipment was placed and <u>DO NOT</u> quote the effect of the explosion on this equipment. For example, the fact that a Centurion tank is here has been published in the press. You may confirm this but do not say what happened to that tank. Lives of your friends in the Royal Armoured Corps or the Royal Australian Armoured Corps may depend on this ... You may describe life at camp as it appeared to you. You may acknowledge the presence of your friends or friends of your friends but do not give a list of the Scientists here and do not detail their tasks ... Do not of course describe the weapon, or the direct effects of the weapon apart from your general impressions of wind, noise, dust, cloud, etc.[59]

During the time X200 was operational, the United Kingdom shipped a total of 463 tonnes of scientific equipment to the site, along with 1142 tonnes of construction materials and motor vehicles and 211 tonnes of rations and accommodation materials, all brought in by air.[60] In addition, another 406 tonnes was brought in by road. Given the short timeframes and the fact that aircraft were landing (initially) on clay, that it was pulled together in time to test atomic weapons is remarkable, and speaks to the urgency of the project.

As plans for the site took shape, so did the wall of secrecy that enveloped a chunk of the Woomera range. As part of this, two peace officers were assigned to Emu. Peace officers were essentially security police employed by the Commonwealth Investigation Service, forerunner of the Commonwealth Police that later formed part of the Australian Federal Police. The request came from the range security officer at Woomera.[61] Peace officers were to live in a tent and do their own cooking, so it was going to take a special kind of person to fulfil the role. As part of their duties, they would be required to maintain radar communication with Woomera and oversee the security of the site. The first to take up his position (on 2 June 1953) was peace officer Sergeant Newall, followed later by a succession of back-up peace officers.[62] Newall's orders were extensive for someone receiving eight shillings per day (assuming he was married – an unmarried officer would only get six shillings and sixpence a day).[63] The seven-days-a-week position was responsible to the range security officer at Woomera to ensure:

(a) That an adequate guard is maintained on all departmental stores and equipment located in the area.
(b) That a record is kept of the movement of all persons passing through the areas, whether NORTH, SOUTH, EAST or WEST.
(c) That only authorised persons move to the WEST.

(d) That the station personnel are inconvenienced as little as possible.
(e) That every assistance is given Range Extension personnel located in the area.
(f) That a close watch is maintained for any speculation or rumour in the area concerning Range activities, including movements of aircraft and convoys.
(g) That as much information as possible is obtained about local inhabitants.
(h) That liaison with local police is made when possible.
(i) That all members of the detachment are able to give advice on roads, NORTH, SOUTH, EAST, etc.
(j) That a nominal roll of station residents in the area is maintained.
(k) That all information possible concerning the NORTH and WEST movement of aboriginals is obtained.
(l) That the views and difficulties of station people in connection with the Range Extension are recorded.
(m) That all members of convoys from site are instructed not to discuss anything in connection with Range Extensions to other than authorised persons.[64]

The plan was to base two peace officers at the site, one working the 6 am to 12.30 pm shift, and the other 12.30 pm to 7 pm. They were instructed to engage any travellers who came near to the site in conversation 'with a view to checking on their story by noticing baggage – for instance – a prospector would not be carrying a tennis racquet'.[65] They could make sure they spoke to travellers by operating a road block to stop anyone passing through. While on-site, they were to keep their tents and surrounding areas spick and span, and had to be neatly dressed when on duty and carry a firearm at all times. Also, '[a] friendly atmosphere should be created with local inhabitants, but this must not be permitted to become too familiar. This means, have occasional meals with the station people if invited, but, don't make it a habit.'[66] Overall, the peace officers were to be 'courteous, respectful and firm'.[67] Plans were made for two peace officers to continue at the site to keep an eye on equipment after the tests, although, as shown in Chapter 10, security at the site was far from watertight.

Penney's offsider at the AWRE, Charles Adams, arrived at X200 in September 1953 and pronounced himself pleased with the work the Australians had done to set it up. 'An excellent job has been made of the camp, beyond what I would have thought possible. No doubt flies and heat will become more of a nuisance later but at present working conditions are pleasant and our expedition is very appreciative of the facilities provided by X200.'[68] A member of Adams' staff, TWJ Redmond, wrote to Titterton telling him what to expect when he stayed at X200:

> Accommodation on site is, as you anticipated, in tents, each of which contains a bed, locker, chair and wash bowl. Apart from the windblown sand you will find your quarters quite comfortable. The charge for messing on site is 1/- [one shilling] per day and payments for incidentals may be made in cash or by cheque drawn on the Commonwealth Bank at Woomera. You would probably find it very convenient to arrange a personal credit at the Woomera Bank.[69]

Both Solandt and Martin advised senior Australian brass that Totem should be a military operation. The Canadian and Australian scientists met with the Australian Defence Committee, comprising chief of the general staff Lieutenant General Sydney Rowell, chief of navy Vice Admiral Sir John Collins, chief of the air staff Air Marshal Sir Donald Hardman and chairman of the Repatriation Commission Fred Chilton, on 9 October 1952. Butement also attended the meeting. The handwritten minutes set out the reasoning for military involvement in these scientific tests. Firstly, the services needed to be able to appreciate exactly what a nuclear war entailed: 'It makes them aware of radiation hazards, safety measures to be taken and produces a sense of reality regarding atomic war'.[70] Also, rather tellingly, the discipline of the services was a necessity. 'Scientists cannot be readily controlled, and one of the less satisfactory features of the MonteBello [sic] test was the haphazard foray of Scientists over the bombed area.'[71] Boots would be needed on the ground, and wayward scientists needed to be contained.

And so, X200 was found and inhabited and put to work as a military installation. Emu Field would never be as busy and established as Maralinga. It was small and temporary and ephemeral, almost a desert mirage. However, the events there changed the course of many lives and have never been completely forgotten. They are carried within the cultural memory of many of those who were affected, even though it remains unknown to the wider Australian population. Emu Field, the first Australian mainland testing site, was where the British, inadvertently or not, experimented on the local Aboriginal populations; the chilling term put to UK atomic weapons chief William Penney during evidence to the Royal Commission was that Australians in range of the tests were 'radiologically expendable', while British people were not.[72] Penney rejected the suggestion. The Australian Government was essentially kept out and did not have a formal mechanism for checking what the British were doing. Indeed, Australia paid at least £134 000 towards the cost of the Emu Field tests, while having no say in how those tests were carried out.[73] The British were free to do with the site what they wished, and that is exactly what they did.

Chapter 2
WHY EMU?

> The decision to use the mainland for atomic tests was made without specific consideration by Australian scientists or others of whether weapons could be safely fired. Consideration was limited to the fact that Emu was a remote location.
>
> Conclusion No. 9 of the Royal Commission into British Nuclear Tests in Australia.

> One can read thousands of pages of documents about the preparation for, and conduct of, the tests without any mention being made of the people whose homeland was being blasted and contaminated. Where there is occasional mention it is more likely to reflect concern about the problems Aborigines might create for those conducting the tests than the problems those conducting the tests created for Aborigines.
>
> Paul Malone & Howard Conkey, *Canberra Times*.

'Why Emu?' is a difficult question to answer. Most obviously, it was within South Australia's Woomera Prohibited Area and therefore already halfway to being a top-secret weapon testing location. The legalities in restricting it as a zone for atomic testing were relatively straightforward. However, below that surface consideration, the question becomes murkier. Emu Field was identified by Len Beadell, with advice and encouragement from the chief scientist Alan Butement. Butement had been involved in establishing the Woomera rocket range and remembered some of the terrain. Beadell was specifically charged with finding a test site that met the criteria of remoteness from human habitation and general suitability for atomic testing. But Emu did not in fact fit these criteria. Human habitations were all around, and the site had no easily accessed water and no road access. It was not suitable for the complexities and risks of atomic testing, or the number of personnel required to oversee such activities.

The site offered plausible reasons for keeping people away, an enormous benefit given the British determination for secrecy. Monte Bello had similar advantages, but its maritime environment and extreme remoteness wore down even the excessively secret British atomic authorities (although they returned briefly in 1956 with the urgent need to test the fission components of the new British thermonuclear weapon before Maralinga was ready). Emu fulfilled its purpose of allowing the British to test highly experimental fission weapons while being able to strictly control access to the site using an abundance of cogent reasons for keeping people away. So the bigger question is, 'Why were the British prepared to put up with extremely poor conditions to ensure that Operation Totem had little or no scrutiny?' We do not have all the answers to that question because of ongoing impenetrable secrecy, but we do have some.

Britain searched all around the world for nuclear test sites, even while engaged in advanced preparations for the Emu Field site, and later Maralinga. The British authorities were never entirely satisfied with Australia as a place to test their weapons – it was too far from home and had a variety of climatic and security issues. The search ranged far and wide, from 'the Shetlands and Hebrides to the Bahamas and Falkland Islands, West Africa, Somaliland, Islands of the Pacific and Indian Ocean', before finally ending up, rather reluctantly, at Emu Field then Maralinga.[1] As Captain FB Lloyd of the Ministry of Supply put it in a letter to the Commonwealth Relations Office, 'Before I report further to the Chiefs of Staff I should be grateful for your comments on whether your Department can suggest any alternative to South Australia which would more clearly meet the stated requirements'.[2]

Australia agreed for a number of reasons including colonial ties, its hopes of creating a market for uranium oxide and its desire for British and American protection in a potential future war. But the most potent reason was that it wanted in on atomic secrets. However, at no point throughout the nuclear tests did Britain share any weapons secrets. Aside from the United States, only Canada was party to such information. Canada had been involved in atomic weapons development from the start and worked closely with Britain and America on the Manhattan Project and in securing uranium supplies so it 'was in a special position'.[3] Australia may have seen itself as special in the eyes of the 'Mother Country', but it really wasn't.

Professor Sir William Penney, AWRE head and architect of the British A-bomb, told the Royal Commission several factors were behind the switch from Monte Bello to Emu Field. 'One of the reasons was that it was clear by the time Hurricane was over that it would not be long before the military people in the United Kingdom would want to put out a weapons test and that needed a lot of area.'[4] The additional space was needed for measurements required by the army, air force and Civil

Defence, aspects of testing he would describe elsewhere as a nuisance from a scientific point of view. Indeed, the military was behind much of the time pressure too, along with senior British politicians who had committed to the nuclear-armed club and wanted to show the United States, even more than the Soviet Union, that Britain was not finished yet in geopolitical terms. 'The second reason was that the scientific people at Monte Bello, Hurricane, complained about the discomforts and not exactly hardships but not very good working conditions'. The idea, he said, was to find 'a range more like Nevada'.[5] Penney had been at home on the Nevada ranges when he was part of the Manhattan Project and, though he could no longer work there, he wanted somewhere similar to test his bomb designs.

Certainly Emu Field had marginally better conditions for personnel than Monte Bello, although its unpredictable winds soon became a problem. Its low rainfall (100–127 millimetres per year, sometimes in one go) was a plus in some ways for the atomic tests, although lack of water for the personnel on-site was, overall, a big minus.[6] The heat was certainly an issue: 'The visibility is excellent at dusk and dawn but may be impaired during the day by heat haze. This may mean that it will be desirable to explode the weapons at night'.[7] Some clueless official at the UK High Explosives Research establishment suggested the Australian military should clear the site completely of flies, a feat, unsurprisingly, not achieved during the short, uncomfortable tenure of the British at Emu.[8]

The Emu Field test site arose abruptly in part due to the shift in the United Kingdom's atomic infrastructure. Two 'atomic piles' (reactors) at Windscale in Cumberland on the country's north-west coast had begun production in 1950 and 1951. They provided high-quality, pure plutonium-239, the expensive raw material for atomic weaponry, with a yearly production target of 90 kilograms. However, when the British moved to develop civilian atomic energy, the Windscale piles were deemed unsuitable, so they drew up plans for new atomic piles at the adjacent Calder Hall. These new piles, known as PIPPAs (a rather forced acronym standing for pressurised pile producing industrial power and plutonium[9]), were designed to create plutonium for both military and civilian uses, making Calder Hall the first place in the world to produce material for civilian atomic energy.[10] Because they had to accommodate two different purposes, the design was different to Windscale's, and the plutonium in the Calder Hall PIPPAs had a higher proportion of plutonium-240, an isotope with different physical properties to plutonium-239.[11] Windscale plutonium contained between 3 and 4.5 per cent plutonium-240, while PIPPAs had around 6 per cent.[12] This made a significant difference to the nuclear fuel the United Kingdom was producing for its atomic arsenal – with unknown consequences for British atomic weapons.

Plutonium-239 is the fuel of choice for atomic weapons. Its rate of spontaneous fission (for the technically minded, about 10 neutrons per second per kilogram) is controllable. On the other hand, according to technology historian CN Hill, plutonium-240

> has a much higher rate of spontaneous fission, producing about 1 000 000 neutrons per second per kilogram. This is highly undesirable in a nuclear weapon, since the neutrons produced by this spontaneous fission can lead to premature detonation of the bomb and hence a reduced yield.[13]

With the shift to Calder Hall plutonium the United Kingdom had to change the way it produced its fission weapons if it wanted to make a lot of atomic weapons in a short timeframe.

Not content to rely on guesswork, the British nuclear authorities needed to test the implications of this change, and fast. The aim of Operation Totem was to see how higher proportions of plutonium-240 in the nuclear fuel would affect performance. No-one really knew the answer. Plutonium-240 was known to make the weapon more unpredictable, and archival documents relating to Emu Field often hint at this. According to a Cabinet document, quoted by Lorna Arnold and Mark Smith, the aim was

> to establish certain minimum characteristics of the fissile material suitable for use in atomic weapons, in order that the existing piles may be operated and future nuclear reactors for the production of fissile material may be designed, to secure the maximum output and lowest cost which is consistent with the efficiency of the weapons.[14]

The British had significantly less than the Americans, but they had to work with what they had. They also had to compromise on their test site, since America was not going to work with them for the foreseeable future. The British cut their cloth accordingly.

Emu was thrown together quickly and with considerable expense and effort, but it was only ever intended to be makeshift. The highly experimental nature of the tests, and the immense political pressure on the test authorities, made the events at Emu highly dangerous and questionable. This site was located in an area long traversed by Aboriginal people, many of whom lived within range of the fallout plumes, and the Totem series put many Aboriginal people in harm's way. Indeed, it could be said that the callous disregard the British displayed towards the local inhabitants of the region had the same effect as if they had overtly sought to use Aboriginal people as experimental subjects.

The ministries of Supply and Defence were central to the British effort to make the bomb. In December 1952, Penney wrote a top-secret letter to bureaucrat FC How at Supply which began, 'There are urgent reasons why we have to explode two more trial atomic weapons as soon as possible'.[15] He did not elaborate on this, but the letter detailed a long list of required actions with an almost impossible deadline of October 1953. These included camps, roads and two steel towers to be provided by the Australian services, the establishment of an executive committee, along the lines of the one that ran Operation Hurricane (HUREX), with Vice Admiral P Brooking (Penney's administrative deputy), a senior Australian services representative in London (approved by the CRO) and officials from the Commonwealth Relations Office (CRO), Treasury, Air Ministry, War Office, Atomic Energy headquarters, Ministry of Supply, Admiralty and Ministry of Transport.[16]

This high-level committee was to meet urgently to swing into action. Penney felt the pressure of getting underway. He would attend the first meeting 'and explain the background more fully'. He also noted that the British high commissioner in Australia had asked the Australians to appoint four people: a brigadier (preferably sapper) to be commandant, a group captain for aircraft matters in Australia, a senior representative of the Australian Department of Supply, and a security officer.[17] These appointees would need to visit Woomera and the proposed test site, then come to London as soon as possible for technical discussions and briefings.[18]

Given these impossible timelines, it is a wonder the tests were held at all. The advance party got to work in February 1953.[19] By 29 November that year, nearly everyone had gone. In those nine months, they created an atomic weapons test site for the two Operation Totem bombs and also set up the site for the five Kittens trigger tests, toxic non-nuclear tests that left a nasty mess behind. Thousands of tonnes of gear were brought in by air and road, often with great difficulty. This was a military operation and, leaving aside the morality of the testing, the dedication, hard work and efficiency of the Australian personnel were extraordinary.

Also in February 1953, a mission of Australians, appointed at Britain's suggestion, visited the AWRE. The group included the head of construction at the X200 site, Brigadier Leonard Lucas, engineer Francis O'Grady, Woomera range superintendent Group Captain Alfred Pither, Dr BG Gates from the Department of Defence and public servant Leo Carter. They visited Fort Halstead, an outstation of the Aldermaston AWRE headquarters, to meet atomic weapons development experts and discuss the difficulties of the X200 site. AWRE scientist Dr LC Tyte noted in his report that the 'natural physical working conditions during the period of the trials were not likely to be intolerable', but they necessitated special requirements to deal with desert temperatures.[20] Flies were identified early as a 'menace' and fly veils were to be provided as protection for the eyes (and presumably for the nerves).[21] Water was another challenging issue. When briefing the Australian mission, the AWRE estimated that about 14 tonnes of distilled water would be required per day, rising to over 36 tonnes per day during the test period itself.[22] Guaranteeing the supply of water was a huge logistical undertaking. At the first meeting, the Australians were told that most travel to and from the site would have to be by air, though some hardy souls would be travelling the long and arduous trip by land. This would take five days from Woomera in vehicles with extra-wide, large-tread tyres. Since only small amounts of gear could be transported this way, the bulk of the equipment would be flown in by military aircraft.[23] The site required 24 vehicles, mostly Land Rovers or Jeeps. Brigadier Lucas said that 10 Land Rovers would be needed for military use.[24]

When the Australian mission asked whether the rough track from Woomera to Emu was likely to become contaminated by the tests, they were assured, rather ambiguously, that 'the contamination, if any, was likely to be light and probably only a nuisance, if at all, for a matter of a few hours. It would, of course, be another question if personnel were to be required to work or live in the vicinity'.[25] Two alternatives were offered for contaminated waste: burial at the site or dumping at sea.[26] The former involved using a bulldozer, which would itself become contaminated, while the latter involved crating the waste in a container and transferring it to 'a dumping organisation'.[27] The AWRE favoured dumping on-site.

Emu Field remains an enigma, functioning on one level as a mere prophecy of what was to come at Maralinga, but without its own distinct identity in Australian atomic test history. In fact, Emu was unique and special, a manifestation of a newly emerging atomic power flexing its muscles and lengthening its stride. In the early 1950s, Britain surfed the wave of patriotism and Cold War fervour. It used the fund of public trust invested in the government from victory in World War II as a cover for its actions, expecting and receiving no thorough scrutiny. The *feel* of the pre-Maralinga tests in Australia is different to that of Maralinga; later, simple patriotism was much less evident.

The Australian population was at first overwhelmingly in favour of assisting Britain, a country held by many to be 'home' and where the Australian head of state, Queen Elizabeth, resided. Elizabeth ascended to the throne at the beginning of the first year of British atomic testing, 1952, and her Coronation was held during the lead-up to Operation Totem in 1953. Her ascension as a young and vigorous monarch set the stage for the rise of a nuclear-armed Britain, and Australia was caught up in the excitement. The mood regarding atomic testing became more complicated from 1956, but at the time of both Hurricane and Totem, the Australian public did not seem fazed by the detonation of atomic weapons on their territory. For its part, the United Kingdom, despite crippling economic constraints, had decided to invest in a nuclear-armed future, and no-one (certainly not the Australian population) needed to know how many corners were cut or people put in harm's way to achieve that ambition. The end was the thing, and the means were nobody's business, besides a small group of insiders. That attitude still obtains today.

The UK-based Totem Executive, known as TOTEX, was led by a rather reluctant Air Marshal Sir Thomas Elmhirst, delaying his plans for a quiet semi-retirement as lieutenant governor of Guernsey. TOTEX was a committee within the Ministry of Supply. Frederick Lindemann wrote to Elmhirst upon his appointment in January 1953, saying:

> I am very pleased indeed that you are willing to accept appointment as Chairman of the Totem Executive. I need hardly say ... your readiness to come forward at the price of such domestic upheaval is appreciated. I am afraid it is going

to be an exacting job but I am sure you will rapidly get on top of it.[28]

Elmhirst was originally employed for between nine and 12 months, although in the event he held the position for just under nine. He was paid a good salary of £2500 per annum, plus generous allowances. The job entailed him signing the Official Secrets Act, signifying his exposure to top-secret information as part of his role.[29]

Elmhirst left his position with a hardly a backward glance on the evening of 15 October 1953, when the mushroom cloud of Totem I was still 24 hours away from fully dissipating.[30] He did ensure, before he departed, that he could still claim expenses; a letter to his contact in the ministry pointed out he still had to finalise his work: 'You kindly said that it would be reasonable for me to be appointed an "honorary consultant", which position would enable me to claim travelling expenses and subsistence while working for the odd day in the Ministry of Supply after the 15th October'.[31]

During his tenure as chairman of TOTEX, Elmhirst also received the pinnacle of aristocratic honours, an invitation to be part of the official party for the Coronation of Queen Elizabeth II at Westminster Abbey on 2 June 1953. As a gold staff officer he was charged with assisting with the seating in the abbey, though the invitation pointedly made it clear that his wife was not entitled to a seat.[32] As loyal troops at the distant atomic outpost raised a beer to the new Elizabethan era, one of the men who had put them there was at the actual event.

As the British and Australian governments firmed up their plans and sent military men to conjure an atomic test site from the clay and dust of the Emu site, they suppressed press reports about the tests for as long as possible. The media was given official information, supplied in carefully worded doses, only three months before the tests were due to take place. Until then they had been fed outright lies, known officially as 'cover stories'. A top-secret meeting at the Ministry of Supply in London in early 1953 charged Penney and Elmhirst with keeping an eye on this. Duncan Sandys, the minister, asked in particular to be kept informed of:

(a) Any proposal to abandon rocket-range 'cover-plan';
(b) Developments concerned with difficulties about Australian aborigines;
(c) Arrangements for satisfying ourselves that Australian security clearances were satisfactory.[33]

Once they put a media campaign into action, the aim was to convey both a sense of patriotic duty and the harmlessness of the tests. A Totem Panel press statement described the site:

> The new forward base which is beginning to take shape is far beyond Woomera across trackless semi-desert. The terrain varies from flat gibber plains [arid land scattered with rocks] covered with grey saltbush, bluebush, and spinifex, and scarred by claypans like the surface of a dead planet, to rolling sand dunes and drifts. Every vehicle or convoy setting out from Woomera is equipped like an expedition with water, stores, petrol, and camping facilities to cover twice the time it is expected to take on the trip. Reconnaissance aircraft are kept in readiness to search for vehicles that might fail to report at a radio-equipped check point on the way. Bogging in the deep sand drifts is the main hazard.[34]

The test site was far more remote than Nevada, or indeed than any other test site in the world at that time. The unaltered claypan with emu claw marks served its purpose as the initial airstrip, more or less. However, pilots had to jam on the brakes when landing, which put a lot of wear and tear on the aircraft and, over time, damaged the landing surface.[35] When heavy loads and a crowded flight roster pushed the capabilities of pilots, aircraft and landing strip to their limits, the Australian military built an all-weather airstrip to ease the congestion.

Unlike Maralinga, built by the civilian company Kwinana, at Emu a mixed force of about 150 RAAF airfield construction personnel, army crew and a few civilian specialists built the base.[36] They used bulldozers and other earthmoving equipment brought overland from Woomera, part of the way on low-loaders and the rest on their own tracks, a hugely laborious process. They even brought in a Centurion tank, one of the target response items required for the tests (the military used target response items to determine the physical damage expected from a bomb). The tank was first shipped to Whyalla in July 1953 and travelled to Mabel Creek on a tank transporter. Nine army personnel shifted this beast of a vehicle, at an estimated cost of £2000.[37]

The men who built Emu had their own newsletter, *Bulldust*, which acted as an outlet for their larrikin humour and, sometimes, their sorrow about being so far from home. Len Beadell – clearly a crowd favourite – was lauded throughout its pages. The editor wrote in August 1953: 'As a typical son of the DEAD HEART, Len has automatically stepped into the editorial circle and it is known that his pen has already caused hilarity throughout Australia per medium of "BULLDUST"'.[38]

While construction crew were largely oblivious to the purposes of their labours, their commitment could not be faulted. The Totem Panel of bureaucrats and military personnel in Australia that oversaw the development of the site claimed that they had 'established construction records that compare with the R.A.A.F.'s greatest wartime airfield construction feats'.[39] They placed Nissen huts on-site to serve as mess huts, recreation huts, offices and workshops. They built gravelled roads through the settlement, as well as roads to the detonation areas. Rows of tents were established to provide sleeping quarters. This hard labour was carried out in unfriendly climatic conditions on unyielding land. In July 1953 the Totem Panel released press materials to journalists that included the following cheerful anecdote:

> 'The men are simply terrific' said a young English scientist who has watched the progress of the base. 'Their enthusiasm is absolutely superb. When they were working under the most primitive conditions I heard the area commandant address them. He said, "Men, I have no good news for you. You'll continue working 10 hours a day seven days a week. Water will be short. There'll be no leave for anybody. I'll try to get amenities, but I can promise only a life of hard toil". Do you know, the troops cheered him.'[40]

While they may have cheered, behind the scenes many of them – or at least their commanding officers – were asking for recompense. The Australian Army officer charged with overseeing construction of the site, Brigadier Lucas, put in a request for special leave

> to compensate for loss of rest days during the period of their attachment [at Emu] ... He has stated that the personnel have been working 7 days a week and will continue to do so until

the conclusion of the Operation. They will be granted leave whilst engaged in this task. For the first 2½ months at least, the conditions under which these personnel have had to work have been extremely severe and it is only recently that it has been possible to provide any reasonable amenities.[41]

The army eventually agreed to provide up to six days of special leave for personnel working at Emu, and the RAAF provided similar consideration, despite some consternation about setting precedents.[42]

At first reveille was at 6 am, and stand down at 6 pm seven days per week.[43] Over time, the seven-days-a-week toil was reduced to six and a half days, with Sunday afternoons off, but working at the site was never easy. The heat was unrelenting and there were few home comforts. However, in June 1953 workers had a patriotic morale boost:

> At one of the loneliest outposts in the British Commonwealth – the site of the forthcoming atomic test – a picked work force of 180 R.A.A.F. and Army men held their own Coronation celebration far out in the Southern Australian desert. Wearing their uniforms for the first time for nearly six months, they took part in a full parade and Coronation service. During the service engineers fired a 21 'gun' salute with expertly timed gelignite charges. Then the desert workers washed the red dust from their throats with a special ration of kegs of beer. The men took part in a sports meeting in the afternoon. In the evening they held a Coronation concert. The star turn was a chain gang under an overseer wielding a whip.[44]

An X200 progress report sent to the Totem Panel on 10 June 1953 described the Coronation events as a credit to the Emu workforce: 'The turn-out and deportment was indicative of the high morale of the Force. In their very limited spare time, the troops had somehow made [the] opportunity to wash and iron and press uniforms for the occasion and very obviously took pride in the result'.[45] The sense of loyalty to the Crown is palpable in the Emu story, as a young and rather glamorous queen took the throne and Britain seemed, on the surface at least, to be reasserting its colonial dominance following the war. The men who worked at X200 knew little about the end goal of their labours; all the same, they were proud to be working in the name of a country they generally saw as part of their Australian national identity.

By June, the workforce was regularly able to watch films and enjoy well-appointed mess rooms and a recreation hut, along with electric light in all tents and a well-stocked canteen, lessening the hardship a little.[46] Games of table tennis, cricket or footy were played during downtime. All accounts point to high morale throughout, possibly because the uncomfortable conditions for the approximately 180 Australian military personnel who were based at Emu during 1953 only had to be endured for a limited period (no more than 10 months). There was a sense of everyone being in it together, even the boffins when they finally arrived. Emu was an egalitarian outpost. The military is well known for providing good food to troops on deployment, so there was that too. Veterans of the Australian nuclear tests often wax lyrical about the generous and tasty rations, even if the rest of their stay was miserable.

Emu was specifically built to test two devices. The two towers built to hold these weapons were located at the ends of the base of an imaginary geographic triangle. The apex of the triangle was the main control and recording point.[47] Totem I was intended to be one-fifth the explosive power of Hurricane, although in the event it was well over one-third. Totem II was intended to be larger than Totem I, but it was not, though it too was larger than anticipated. These errors highlight the inaccuracy of the British predictions of explosive yield.

Britain was nevertheless ahead of many other nations in the early days of atomic weaponry, despite its significant knowledge gaps. The first British atomic bomb test, Operation Hurricane in 1952, had proved William Penney's innovative Blue Danube design, a bomb with some features that made it superior to the weapon dropped on Nagasaki by the Americans in August 1945. Penney was acknowledged even by his critics as a genius when it came to atomic weapons, and his political masters were seemingly in awe of his intellectual and practical capacities. He led the British Mission of scientists to the Manhattan Project and was recognised for his calm brain and steady disposition, and the academic rigour that came from his two PhDs and his civilian role as a mathematical physicist. He was first drawn into secret defence work in Tube Alloys, the deliberately enigmatic name for the clandestine British project that preceded the Manhattan Project, where he demonstrated a highly valued combination of trustworthiness and creativity. He created both the United Kingdom's A-bomb, and its hydrogen bomb (H-bomb). Underneath his positive demeanour, he was often conflicted, seeing himself as a bookish don, not a warmonger. In the late 1950s he bowed out of nuclear weapons design and testing, and later advocated disarmament. Blue Danube went out of production in 1958, when American designs entered the UK arsenal.[48]

While Operation Hurricane had satisfied a particularly British interest in atomic warfare conducted in the naval environment, in the early 1950s the main game was atomic weapons dropped from aircraft. Emu Field was identified before Operation Hurricane, as future testing needs unfurled in front of the small group of British insiders who plotted the development of the nation's atomic arsenal. Indeed, Penney's trip to the Emu claypan en route to Monte Bello before Operation Hurricane alarmed Sir Frederick Shedden, the secretary of the Australian Department of Defence. He had seen the correspondence during August 1952 about the visit, but there was no formal request to the Australian Government to investigate the feasibility of atomic testing in the South Australian hinterland.[49] When Sir John Cockcroft finally sent an official request to Prime Minister Menzies in September, the arrangements had effectively been made. The British were set upon a course of action, dragging a largely ignorant and uncomprehending Australian Government well behind them. Physicist Dr John Symonds, who was commissioned in 1984 by the Department of Resources and Energy to prepare a history of the British tests for the Royal Commission, quoted Cockcroft's note to Sir Roger Makins about his communication with Menzies: 'I saw the Prime Minister today and asked him whether we might have facilities on Woomera Range for further tests if desired. He is agreeable in principle and to Penney making reconnaissance for possible sites'.[50]

The limits of the site allowed for only 300 people at a time, including 60 scientific and technical staff, 300 VIPs or other 'extra mural representatives' and the rest administrative and military personnel.[51] The claypan airstrip, and later its all-weather clone, was in regular use from the start: 'initial requirement is for 40 sorties by Bristol Freighter during period of one month beginning 14th February. This would involve approximately 160 flying hours. R.A.A.F. is hard pressed to meet demands for transport aircraft, and Dakota type aircraft are not suitable for loads expected in operation'.[52]

Well before the tests, the British knew that Aboriginal people lived in the vicinity of Emu Field. However, they shifted responsibility onto the Australian Government and remained unmoved and incurious about the people who made central Australia their home. The assistant under-secretary for Commonwealth relations and later British deputy high commissioner to Australia, Neil Pritchard, wrote to Ben Cockram in the British High Commission in Australia on 2 March 1953, saying: 'There is one aspect of the coming Totem test which may cause us some difficulty from the political and publicity point of view'.[53] This letter sets out the broad attitude of the British authorities at the time:

> We are told that the site for the test is in fact far removed from normal human habitation but that the general area is close to territory peopled by aboriginal inhabitants. We gather that there may well be agitation at some later stage about the possibility that the tests may place in jeopardy some of these aborigines and if public comment of this type is aroused we must anticipate alarm on the part of well-intentioned persons and societies, in the United Kingdom as well as in Australia, which might lead to Parliamentary questions ... Needless to say we are assuming that no danger exists of the tests resulting in the death or injury of any aboriginal inhabitants. But it is for the Australians to make sure of this and to take the lead in handling any public controversy.[54]

In answer to questions here probably our reply should be that the Australian Government have satisfied themselves that no aborigine will come to harm. Whilst there is presumably no great urgency since there is no prospect of an announcement about the test for some time, we should be grateful if you could open up this matter at your convenience with the Australians and let us know broadly what their arrangements are and how they would like us to reply to any enquiries we may get later on from the Aborigines Protection Society et. al.[55]

In fact, the Australian Government took little more notice of the issue than the British did other than to assign NPO Walter MacDougall to keep Aboriginal people away from the new Emu site. In August 1953, the secretary of the Aborigines Protection Board in Adelaide sent a memorandum to MacDougall, saying:

I have written to the Department of Native Affairs, Perth, pointing out that it is undesirable for any natives to travel towards or in the vicinity of the L.R.W.E. Woomera and made the request that you be advised by the Cundeelee Mission authorities of any movement of natives towards the prohibited area.[56]

MacDougall was certainly knowledgeable about the people and the land, but he could not possibly stop all movement of people over such a vast area, especially during the dingo pup hunting season, which coincided with the tests.

Totem was planned quickly and in complete secrecy. While the various Machiavellian wheels-within-wheels in the British government and civil service moved the atomic armament of Britain forward in all sorts of secret ways, a more practical and pragmatic approach was needed in the interface with the Australians. The Totem Panel was made up of senior civil servants and senior officers of the three armed services, together with a scientific representative from the UK Ministry of Supply and chaired by Major-General JES (Jack) Stevens. Later to become Sir Jack, Stevens was a major-general during World War II, commanding the Sixth Division in New Guinea, known as the 'Fighting Sixth'. After the war, he was appointed as the secretary of the Australian Department of Supply and at the time of Totem was chairman of the new Australian Atomic Energy Commission.[57] (His son, Duncan Stevens, was captain of the *Voyager* and among the 82 lives lost when his ship sank after colliding with HMAS *Melbourne* in February 1964.[58])

Australia was hardly oblivious to the cataclysmic new weapons created by the Manhattan Project. Celebrated Australian physicist Mark Oliphant had provided some insights to the Australian Government during the war, before the new weapons were demonstrated at Hiroshima and Nagasaki. In 1949 the Soviet Union revealed via an unexpected atomic weapons test that it had mastered the technology, courtesy of several spy-physicists from the Manhattan Project who passed on atomic secrets. In that same year, Australia's Defence Research and Development Policy Committee summed up its own evolution in light of these post-Hiroshima events. 'A substantial proportion of Australian defence research is already committed to matters which affect the efficiency of delivering and countering atomic attack', according to the minutes of a meeting held on 7 December 1949 and chaired by Professor Leslie Martin.[59] The notes listed these research activities as: 'improvements in air navigation, fighter control, and the provision of range facilities in support of guided weapon developments. It remains to be considered whether anything should be done in Australia to increase the range of production of atomic weapons by friendly countries'.[60] In the late 1940s Australia was prepared to answer the call, if it came, to assist in building atomic weaponry.

In July 1954, the UK Cabinet made a strategic decision to pursue thermonuclear weaponry, hydrogen bombs, based on fusion technology.[61] While fission devices rely on the splitting of heavy elements (such as uranium or plutonium) into lighter elements, fusion works in the opposite direction and forcefully fuses light elements, usually forms of hydrogen.[62] These weapons generally require a fission explosion to produce the enormous pressure and heat needed for fusion, so H-bombs are also known as thermonuclear weapons and their yield is measured in megatons (1000 kilotons). For a fission bomb to create such a huge blast, it would have to be enormous.[63] The British fission weapons tested in Australia never exceeded one-tenth of a megaton, and most were much smaller.

In both Operation Hurricane and Operation Totem, the technology was entirely fission based, 'old' technology (dating back to the Trinity test in 1945 and the bombs dropped on Hiroshima and Nagasaki). The tests also investigated radiation effects and a wealth of other matters related to atomic weapons but the British quickly moved on when the technology tested at Emu fell by the wayside. The decision to pursue the H-bomb marks a significant shift in the two and a half years between Emu Field in 1953 and the later Monte Bello and Maralinga tests in 1956 and 1957. British priorities, the politics around atomic weaponry and indeed the technology itself fundamentally changed over this period. Despite some contemporary and more recent conjecture, no hydrogen bomb was exploded on Australian territory. However, operations Mosaic and Antler were connected to the H-bombs that Britain later tested in the Pacific, at Christmas Island (now known as Kiritimati, part of Kiribati). These tests came at great expense to Pacific Islanders and their environment. Operation Totem stands as an anomaly, a largely useless dead end that had the worst consequences for those with the least power.

Australia envisaged an atomic future for itself, and expected to bootstrap off its agreements with Britain to test its own atomic weapons one day. The young Australian scientists who worked with Titterton on measurements at Totem were part of this program of gearing up to be equipped for both civilian and defence uses of atomic science. The Defence Research and Development Policy committee laid out plans for how this would develop: 'the Australian Atomic Energy Commission [will] arrange – so far as security permits – for Universities, C.S.I.R.O. [Commonwealth Scientific and Industrial Research Organisation] and other suitable Governmental or private organizations to undertake research work in the fields of nuclear physics, metallurgy, analytical chemistry, chemical and mechanical engineering'.[64] Initially Australia would rely on selling the raw material, uranium oxide, to the United Kingdom where it would be converted into the metal fuel elements for use in a putative Australian reactor. Over time it planned to develop sufficient expertise and infrastructure to become self-sufficient. Emu Field was part of a plan by Australia, entirely unrealised, to be a nuclear player. Instead, all it would have to show down the track was an expensive and shameful aftermath.

Chapter 3

TOWARDS A BRITISH 'AUSTERITY BOMB'

> Although the acquisition of the bomb in 1952 had reaffirmed Whitehall's belief in Britain's Great Power status, such achievements masked the straitened circumstances of British research establishments dogged by shortages of materials and staff at every stage.
>
> Martin Theaker, 'Being Nuclear on a Budget'.

Behind closed doors in Whitehall, a secret Cabinet committee made the momentous decision that would resound through Australia for generations. On 8 January 1947, the committee known as GEN.163 met at 10 Downing Street and decided that Britain would proceed with atomic weapons research and development. GEN.163 included Labour Prime Minister Clement Attlee and his deputy, Herbert Morrison, Foreign Minister Ernest Bevin, Minister of Defence AV Alexander, Minister of Supply John Wilmot, and Controller of Atomic Energy Lord Portal. Portal declared that it would take at least three years of work to solve basic difficulties in nuclear physics and to engineer the bomb mechanism. All agreed that the right person for this job was Dr William Penney, long regarded as Britain's best atomic weapons mind. Penney learned of GEN.163's decision on 12 May 1947. After he had been banished from working with the Americans by the McMahon Act of 1946, he had returned to the United Kingdom with invaluable experience, full of plans for a British atomic weapon.[1] Penney loved the intellectual challenge of the task.

Like many major events in history, Operation Totem had a long and tangled causal chain. The trajectory was set when the United States passed its infamous McMahon Act, which banned co-operation in the development of atomic weapons. Pioneering UK-based physicists were on their own. The GEN.163 committee took its fateful decision to develop its own atomic weapons, knowing it would face opposition within government ranks and place a considerable burden on a nearly bankrupt country.[2] The reasons for the decision were complex, but national pride (some would say hubris) was certainly a factor. The country also determined to pursue civilian atomic energy as a natural partner to atomic weapons development.

Two key positions to realise this ambition were created well ahead of the decision itself. On 29 January 1946, Attlee announced in the House of Commons the appointments of Dr John (later Sir John) Cockcroft and Lord Portal (Charles Portal, 1st Viscount Portal of Hungerford). Cockroft, an eminent physicist and Tube Alloys and Manhattan Project insider, would head up the atomic research establishment at Harwell.[3] Portal, chief of the air staff during World War II, became controller of atomic energy, responsible for overseeing the development of UK atomic energy and weaponry (although the United Kingdom's ambition to produce an atomic weapon was not yet formally decided).

Attlee had sounded out Portal before the end of 1945, telling him, as Portal recalled, 'that my chief task would be to decide when engineering projects should be started'.[4] Portal was specifically charged with ensuring that the scientists and the engineers could work together effectively: no simple matter.[5] During the Manhattan Project, the British Mission of scientists and technicians sent over the Atlantic had been excluded from understanding many of the associated technical processes. Britain needed to fill this knowledge gap before it could make its own weapons. From this perspective, the McMahon Act, despite the resentment it generated, concentrated the minds of British scientists and engineers. British nuclear engineer Sir Leonard Owen believed 'the McMahon Act was probably one of the best things that happened to the technologists of the British Atomic Energy Project as it made us work and think for ourselves along independent lines'.[6] The pathway set by the McMahon Act took the British all the way to outback Australia.

An important character in this part of the story is Frederick Lindemann, Britain's most influential booster of atomic weaponry. Frederick Lindemann, Lord Cherwell, the 1st (and last) Viscount Cherwell, was a physicist and a controversial figure. His mother, Olga, was an American with Russian and English heritage, and his wealthy engineer father, Adolph Friedrich Lindemann, was a British citizen who had left Germany as a young man. Olga had decided to 'take the waters' at Baden-Baden at the end of her pregnancy, so Lindemann had been born in Germany, a fact that caused him difficulties with his passport throughout his life.[7] He also undertook most of his schooling and university studies in Germany, and had a German accent.[8] He held problematic views on the discredited theory of eugenics, and was generally haughty and dislikeable.[9] His close, though sometimes prickly, relationship with Churchill was often centred on the great existential questions of nuclear weaponry and warfare. Churchill had an inordinate fondness for Lindemann, and was said to have proclaimed in parliament in response to a critical comment about him 'Love me, love my dog, and if you don't love my dog you damn well can't love me'.[10] Their friendship was in some ways that of opposites. Churchill was known for his appreciation of cigars, fine (carnivorous) dining and the very best whisky, while Lindemann was a non-smoking vegetarian teetotaller. Despite this, the pair dined together frequently.[11]

Interestingly, Lindemann, although a Tory through and through, was a member of Portal's atomic bomb technical committee formed under Labour.[12] When Churchill came back to power in 1951, Lindemann was appointed paymaster-general, a position that entitled him to sit in Cabinet. Churchill, rather unexpectedly, gave this position an explicit atomic theme, making Lindemann 'responsible for advising the prime minister and the Cabinet on policy matters relating to atomic energy and for taking all ministerial action required in regard to research, development and production'.[13] Lindemann, an insider who knew many of the technicalities of the British bomb, was primed to become its most ardent proponent in government.

Portal received his first bomb-making orders on 9 August 1946 (five months before the official decision to build it), in the form of an Air Staff Requirement notice that required him to oversee: 'the development of a bomb [employing] the principle of nuclear fission'.[14] The weapon was not to exceed 4.5 tonnes, be under 736 centimetres in length and 152.4 centimetres in diameter, and was to have flip-out fins that would extend after release and a classic aerodynamic bomb shape that would fit snugly into the bomb-bays of a new British aircraft on the drawing board called the V bombers, which came into service in 1955. The device was to be suitable for release from 20 000 to 50 000 feet at speeds of between 150 and 500 knots.[15] The United Kingdom was gearing up its new air fleet for the atomic age.

The next Air Staff Requirement notice, dated 7 November 1946, informed Portal that V bombers – 70 Valiants, 15 Victors and 15 Vulcans – would carry the bomb. These new aircraft would be capable of carrying a 4.5 tonne bomb to a target 1500 nautical miles from base, at a cruising height between 35 000 and 50 000 feet. The aircraft would rely 'on speed, height and manoeuvrability against interception', and would be equipped with state-of-the-art early warning devices and radar countermeasures to deflect a beam on which a controlled weapon might be launched.[16]

Lord Portal ran a small committee with a large quotient of brain power. It included pioneering nuclear physicist Sir John Cockcroft, the engineer Sir Christopher (later Lord) Hinton, physicist and former Maud Committee member Mr (later Sir) Michael Perrin, as well as Penney. As Portal later put it, 'Cockcroft was responsible for science, Hinton for engineering, Perrin for assisting me at headquarters, and Penney ... for weapon development'.[17] All these men were considered brilliant in their roles. They navigated the shallows of British bureaucracy like masters and commanders. Portal believed that it would have been an 'expensive fiasco' if not for their competence.[18] One could perhaps have wished for less scientific and engineering competence, given the damage done by British nuclear testing.

The British nuclear weapons pioneer and Nobel laureate in physics Sir James Chadwick guided Portal in the early days of his tenure. Chadwick had been influential in the British Mission to the Manhattan Project and was a member of the

British Advisory Committee on Atomic Energy, set up at the end of the war. He felt that Portal was a kindred spirit, noting in a handwritten letter to him in 1966:

> I remember so well our first meeting, when you invited me to lunch in The Travellers Club. For the first time in London I met a man who appreciated the implications of the nuclear bomb in future warfare and who was prepared to do something about it. It was a great encouragement to me.[19]

Few people were better placed than Chadwick to advise Portal on how to put together a British bomb program on the cheap.

Even before the final political decision was taken to pursue atomic weaponry, Chadwick and Portal were proceeding as if it were inevitable. In a personal letter to Portal in April 1946, around the time he was appointed British scientific advisor to the new United Nations Atomic Energy Commission, Chadwick set out the case of scaling up British production facilities to make weapons-grade plutonium: 'If we think that military security and defence depends on our having atomic bombs, then we must make plutonium. It is perfectly clear that at the present time plutonium is much superior to [uranium-]235 for this purpose.'[20] Plutonium had powered the weapon dropped on Nagasaki, as opposed to the one dropped on Hiroshima, which was powered by uranium. The physical properties of plutonium-239 make it superior for atomic weaponry.

Portal, coming off a low base, needed logistical skills to apportion effort in the fast assembly of a bomb-making enterprise. He was allocated 1225 tonnes of uranium oxide, expected to yield 817 tonnes of uranium for use in the Windscale reactor (and Calder Hall when it came online in

1956), to make the weapons-grade plutonium.[21] Chadwick calculated this would produce 180 kilograms of plutonium and 'make 30 bombs of Nagasaki type, or 36 bombs of the new type [which became known as Blue Danube] which has not yet been tested'.[22] Chadwick wanted to know if the Chiefs of Staff considered this enough to fend off the Soviet Union. If not, then Britain lacked the raw materials for its atomic weapons ambitions. What could be done?

Chadwick then described a way to maximise the bomb-making potential from a finite amount of uranium ore. The new Calder Hall pile would make plutonium with a higher proportion of plutonium-240, potentially creating instability in the atomic fuel:

> The production per ton can be increased by leaving the metal longer in the pile. This has two disadvantages: (1) some of the slugs [lumps of plutonium] may develop bulges, (2) the [plutonium-]240 concentration is increased, giving a greater probability of pre-detonation of the bomb and a somewhat lower efficiency on the average. In this way we might get 360 kg. of plutonium.[23]

This would enable production of nearly twice as many bombs, 50 to 60, while still leaving about 60 kilograms of plutonium for experimental purposes. Chadwick argued that 360 kilograms with up to 4 per cent plutonium-240 was of greater military value than 180 kilograms with less than 2 per cent of 240. 'The greater number of bombs more than compensates for the unreliability.'[24] The goal here was clear: use whatever means necessary to produce nuclear fuel. Could atomic weapons be made from the plutonium-240-rich material churned out of the new reactors? Would such bombs work as intended, or would they explode in an uncontrolled way? Operation Totem was devised to answer these questions.

In further correspondence, Chadwick elaborated on the plutonium-240 issue:

> The longer the metal is left in the pile the more plutonium 240 accumulates in the plutonium 239, since it is made from 239 by the simple process of neutron capture. The 240 tends to cause pre-detonation in the bomb so that the concentration should be kept below a limit which corresponds to a time of about 200 to 250 days.[25]

The new UK plutonium production methods were treading a fine line between productivity and risk. This may explain why the Blue Danube service weapons that were later deployed to the Royal Air Force (RAF) did not contain the plutonium-only fuel tested in Australia. Instead, they had a composite plutonium-239 and uranium-235 core, which was less likely to pre-detonate and was also cheaper to produce.[26]

And so a leading expert advised the British atomic authorities to use lesser quality plutonium to increase the numbers of bombs. The tests at Emu Field expressly tested the feasibility of nuclear fuel with a higher proportion of plutonium-240, to create an atomic arsenal on the cheap. Chadwick continued in his advice to Portal:

> According to a statement made to me by General [Leslie] Groves [US military head of the Manhattan Project], it would be possible to increase the yield per ton by putting back into the pile some of the uranium which has been used in the pile and depleted in 235. This has not been attempted

in the U.S. since their extraction process does not separate the uranium from the fission products.[27]

Working the plutonium in this way could provide, at most, '50–60 bombs of a lower degree of reliability than the Nagasaki bomb'.[28] Chadwick conjectured that Britain would need 50 bombs and even then, 'I should still be somewhat uneasy'.[29] He was in favour of pushing ahead with the 'austerity bomb', even though the end product was likely to be flawed:

> If we take this number as 50 then we see that we can provide it from our allocation on our present plans only by pushing the yield of plutonium to an extent which entails a certain degree of unpredictability in the action of any single bomb, although with little loss of efficiency over the average of a large number, and that we shall have at most a residue of about 100 kg. for experimental and development purposes.[30]

In working within its resource constraints, Britain decided to make drastic compromises. No doubt its scientists and engineers looked with envy at the Americans.

On that winter's day in January 1947, the GEN.163 meeting resolved to keep the bomb-making activities secret for a period of time. Supply Minister Wilmot sent a memorandum to Portal that nominated Penney to have carriage of the nation's top-secret nuclear ambitions. He said:

> The Chief Superintendent of Armament Research (Dr Penney) has been intimately concerned in the recent American trials and knows more than any other British scientist about the secrets of the American bomb. He has the

facilities for the necessary research and development which could be 'camouflaged' as 'Basic High Explosive Research' (a subject for which he is actually responsible but on which no work is in fact being done). His responsibilities are at present to the Army side of the Ministry of Supply, but by special arrangements with the head of the Department he could be made responsible also to me for this.[31]

Britain resented the way America ruthlessly pursued its own nuclear agenda with little regard for Britain's original contribution or its future plans. These tensions had a long history. The Quebec Agreement, concluded in August 1943, outlined terms for co-operation on atomic weapons and energy, after difficult negotiations between Winston Churchill and Franklin Roosevelt.

The Quebec Agreement gave the United Kingdom a veto over launching a nuclear strike. It expressly forbade the two powers using the weapons against each other, or against a third country without mutual agreement. But the agreement also foresaw the economic implications of this new technology and allowed the United States to make all the decisions over civilian use:

in view of the heavy burden of production falling upon the United States as a result of the wise division of war effort, the British Government recognise that any post-war advantages of an industrial or commercial character shall be dealt with as between the United States and Great Britain on terms to be specified by the President of the United States to the Prime Minister of Great Britain.[32]

Simmering resentments towards the United States were many. Chadwick was among those in the British camp who thought that the United States took too much credit for the development of the A-bomb, and had used mechanisms such as the Quebec Agreement to steamroller Britain. George Strauss, a long-serving Labour Party politician and minister of Supply between 1947 and 1951, asserted in his unpublished autobiography that the US government had (cynically, he implied) turned the Quebec Agreement into a mechanism to stymie Britain's nuclear ambitions, both military and civilian. He claimed that the US government had opportunistically interpreted clauses to make it easier to prevent Britain from building its own nuclear plant:

> Attlee losing patience, cabled Truman asking for the release to us of such information as would enable us to build a nuclear plant in Britain. Truman, fearing that this would lead to a head-on conflict with Congress, refused. And he threatened that if we proceeded unilaterally, the setting up of [the United Nations] might be jeopardized. Attlee replied that the United States Government was misinterpreting the Quebec Agreement to the grave detriment of British interests.[33]

Strauss maintained that this damaged relations between the two countries. Britain made the decision to proceed despite significant tensions with its major ally.

Strauss was a member of the Cabinet Defence Sub-Committee, chaired by Attlee, which included the Chiefs of Staff. During a meeting of the committee in 1948, the chancellor of the exchequer, Sir Richard Stafford-Cripps, asked about the cost of atomic weapons. The meeting was told that the cost could run into the hundreds of millions of pounds and the first bomb would take about five years to build. Strauss further noted,

> And what effect would a decision by us to go ahead alone have on our relations with the United States Government? We were told it would probably exacerbate the prevailing coldness. It was clear that the United States didn't want us to have bombs of our own as this would deny them their monopoly ownership, outside the U.S.S.R., of this decisive weapon. And there was the possibility that if there were a production plant in Britain, the Russians might capture it together with our stock of fissile material![34]

According to Strauss, Attlee summed up by saying that Britain could not rely on the United States to help Britain at all by supplying bombs or information should a war threaten in Europe. They were alone, after the drama and ultimate success of the Manhattan Project to which they had contributed so much. Therefore, Strauss said,

> My Ministry was directed to proceed with the bomb's development as speedily as possible and to spend on it whatever was necessary. This was then thought to be about £100 000 000 during the first five years. That figure was not identifiable in the Ministry's published estimates, but included under other headings.[35]

The budget was huge by UK austerity standards, but dwarfed by the US$2 billion price tag for the Manhattan Project.[36] Much of that hidden UK budget expenditure would fund operations Hurricane and Totem.

Pressure on the UK decision-makers intensified on 3 September 1949, Strauss recounted, when 'to the surprise of my Ministry, Truman announced that an atomic bomb had been exploded in the Soviet Union. We had been assured by our experts that it would be many years before this could happen'.[37] The bomb-makers needed to get atomic weapons tested and deployed.

The political architect of Britain's atomic bomb, Clement Attlee, was ousted from office on 26 October 1951 and replaced by Churchill, aged nearly 76 and unwell. Churchill, something of a science enthusiast, had been egged on by Lindemann during the war. Now in his second stint as prime minister, with Lindemann still at his side, Churchill's enthusiasm returned.[38] Indeed, without Lindemann, Churchill might well have been content to revel in his past success. His declining vigour had an effect on many things in his second term, including taking on onerous new projects. But Lindemann fired up his friend's spirit to make Churchill 'the leader who truly defined Britain's nuclear age', the country's atomic prime minister, ahead of Labour's Clement Attlee who had initiated the atomic bomb program.[39]

Before his re-election, Churchill was a critic of Attlee's nuclear program, carping that it was too slow and generally inadequate.[40] But despite his sniping as leader of the opposition, once he returned to power Churchill was surprised by the progress of the Attlee government and impressed that it had spent so much covert money on the bomb-making venture, with the huge expenses hidden in the national accounts.[41] This ability of the British government to bury the nuclear test program and share only bland bits of information has continued to the present.

Churchill, though, was capricious. When he learned the

extent of planning for the first test at Monte Bello, he originally stated – to Lindemann's horror – that he did not think that Britain needed to manufacture bombs itself, merely that it should develop the know-how, 'the art rather than the article'.[42] Part of his concern was the £100 million price tag, a sum that would surely rattle any incoming prime minister of a country impoverished by war. However, national pride and fascination for atomic technology, along with some frantic lobbying from Lindemann, soon turned him around and he gave the green light for Attlee's carefully laid plans to proceed.[43]

Historian Margaret Gowing points out it was inherently 'stupid' that the British had to test their weapon in Australia, given their scientists were already familiar with the well-developed US test sites.[44] However, backchannel negotiations between Britain and the United States in the dying days of the Truman administration collapsed when the United States imposed a humiliating set of onerous conditions on Britain. Unprepared to accept America's draconian (not to say insulting) conditions, Britain defiantly put national pride first. The decision had risks; the British had a gnawing fear, felt right up to the detonation of Operation Hurricane, that the British bomb would not work. An unsuccessful detonation, Lindemann wrote in a minute paper to Churchill, quoted by Gowing and Arnold, 'would be disastrous if this happens in full view of all the United States newspapers'.[45] What a bad look. How inimical to Britain's geopolitical aims.

Initially, Britain pursued atomic weapons to bolster the conventional armaments of the nation. Both Attlee and Churchill saw fission weapons as part of the arsenal of conventional warfare. After all, A-bombs could be selectively aimed at military targets and their destructive properties could be geographically contained, more or less. The face of warfare was truly changed with the advent of thermonuclear weapons, which were, for Churchill, a source of considerable terror (and fascination).[46] H-bombs were not designed for military targets. They were aimed at wiping out enormous civilian populations in one hit. 'Subjectively', as historian Kevin Ruane put it, 'the H-bomb scared [Churchill] half to death. Objectively, he knew it was a necessary evil both to improve Britain's national security and to buttress its flagging great power status'.[47] Churchill, and his Conservative Party successors Anthony Eden and Harold Macmillan, saw mastery of these horror weapons as a way of Britain returning to its favoured status with America. As a bonus, Duncan Sandys' famous Defence White Paper of 1957 outlined that it would allow the United Kingdom to cut back on the number of troops in its armed forces by the end of the 1950s, when it would be nuclear armed to the hilt.[48]

At first, Britain focused on airborne weapons. The HEROD Committee had been formed in 1948, with the express purpose of introducing the atomic weapon into the RAF, and dealing with all associated issues and problems.[49] Even before Hurricane, Penney was concerned to get the RAF up to speed on atomic weapons. In a top-secret note he told Air Chief Marshal Sir Ralph Cochrane:

> My philosophy is that the R.A.F. has handled air-craft for a long time, and can fly Valiants as soon as they come off the production line. But the R.A.F. has not yet handled atomic weapons. Therefore, we must get some bombs to the R.A.F. at the earliest possible moment, so that the handling and servicing can be practiced and fully worked out.[50]

Penney believed this would boost the morale of the RAF in particular, and the country in general, though he added a note of caution:

> It would be optimistic to assume that we shall have a completely proven weapon in service before mid-1954. This, however, does not mean that H.E.R. [High Explosives Research] will not be handing over weapons to R.A.F. until then. We still expect to start delivery of the first about July next year [1953], although these weapons only have a limited clearance. There are many handling and servicing problems and the sooner the R.A.F. gets practice the better.[51]

The plan was to deploy a small number of initial weapons, while continuing to test and modify the design. Essentially the British started deploying a weapon that was not fully tested. Operation Totem was the next step to make Blue Danube suitable for full deployment.

So while the success of Operation Hurricane at Monte Bello in 1952 led to the initial deployment of a small number of Blue Danube bombs almost immediately, the bomb needed refinement before it could be fully deployed. In addition, economic and resource constraints meant they needed to be made cheaply. A weapon designed to be exploded on the ground is different to one to be dropped from an aeroplane – in its casing, fusing, arming and storing.[52] This meant negotiating a steep learning curve between the military and the scientists:

> It was different from the usual military requirements because they had no experience in the field. There was a great deal of give and take between the military technical people and us.

It came down to their taking what they could get. They had to have requirements on the operational side.[53]

The RAF geared up to deploy the bomb as part of the new generation of national defence (though dismissing the name Blue Danube as 'a Ministry of Supply name'[54] they instead called it 'Small Boy', despite the confusion this caused with Tall Boy, the wartime British conventional bomb). At a top-secret meeting of the HEROD Committee in July 1953, with Operation Totem just four months away, the go-ahead was given to ready 60 atomic bombs for immediate assembly, in time for the arrival of the V bombers.[55] These plans were being made in the context of a belief that atomic warfare was imminent. The agenda of the HEROD Committee meeting of July 1953 put it starkly:

> The Chiefs of Staff, in their review of Global Strategy, envisaged an initial period of all-out offensive on the outbreak of war, during which both sides would attempt to alter the balance of strengths so as to make the final outcome of the war certain. Clearly in order to meet the Russian threat we should be prepared to launch a maximum effort initial assault against his offensive air, ground and sea power immediately on the outbreak of war.[56]

The United Kingdom was preparing for an atomic war. In 1953 it built a secret atomic bunker 40 metres underground in Scotland, designed to house a government command centre in the event of an atomic strike. The anticipated enemy was the Soviet Union, armed with atomic bombs and soon to have thermonuclear weapons. Military planning determined that

Britain would need to attack 45 enemy airfields in the initial assault, along with 'other priority targets' such as Soviet naval bases and sea communications.[57] The sheer horror of an atomic war pushes through the bland bureaucratic wording:

> In the light of Global Strategy it appears probable that each opposing air force will attempt to destroy the other on its bases. It thus seems possible that on the return of our atomic bombers from the initial assault they may find their airfields have been subjected to heavy attack. Whilst diversion to other emergency airfields would permit the aircraft themselves to be prepared for a second assault, special arrangements may have to be made for the provision of bombs at these emergency bases.[58]

Unstated but implicit is the devastation of the civilian populations around those airfields.

The success of Hurricane, and the fact that it made the United States sit up and take notice of the nation's newly revealed atomic expertise, spurred the British government onwards. In December 1952, the Defence Committee considered proposals from Lindemann to increase production of plutonium and to authorise further atomic tests in Australia.[59] Churchill first wanted to consult with the new American president, Dwight Eisenhower, about whether increasing British production of both weapons-grade and reactor-grade plutonium might frighten the horses across the Atlantic. He journeyed over that ocean in January 1953 to begin a long dialogue with Eisenhower about their atomic relationship.[60] While Eisenhower did not initially know about the Quebec Agreement, and was unenthusiastic about reviving it when briefed, he did tell Churchill that he was opposed to the

McMahon Act and would work to restore nuclear co-operation between their two nations. As Ruane put it, Eisenhower 'was personally convinced that US national security stood to gain from closer atomic associations with the United Kingdom. Against this, the President-elect made clear that there was not the remotest chance of a resuscitated Quebec agreement'.[61]

Churchill met with Eisenhower despite Lindemann arguing that Britain was better off without the United States and should chart its own path. Lindemann believed that realising its atomic ambitions by co-opting the Commonwealth would give Britain more clout with the Americans, particularly if they managed to monopolise a decent proportion of the world's uranium. He was wedded to the idea of Britain exploiting nuclear energy for both military and civilian uses. As he eloquently put it in a secret letter to Churchill in December 1952:

> Our prosperity in the Victorian era was due largely to the men who had the imagination to put and keep England ahead for 60 or 80 years in the use of steam power for industrial purposes. It is quite likely – especially in view of the rising price and slowly dwindling reserves of British coal – that prosperity in the coming century may depend upon learning how to exploit the energy latent in uranium (1 lb equivalent to 1000 tons of coal). No one, of course, knows exactly how and when these developments will come; but nobody knew 150 years ago the part the steam engine would play.[62]

Britain had to embrace a nuclear future if it wanted to prosper. He told Churchill that the cost of developing atomic weaponry and civilian power was no more than the country would pay to stick with existing means, but they would ultimately pay a higher price for refusing to modernise. '[W]ithout plutonium and active uranium no progress in industrial applications can be made.'[63] His fervour – indeed, sometimes, his fanaticism – shines through much of his correspondence.

Lindemann's vision for a nuclear United Kingdom had a way to go. Penney knew better than anyone how much was still to be done before the United Kingdom reached full atomic weapons proficiency. Once he had chosen Emu Field, he began preparing for a hectic testing schedule, limited by a narrow climatic window. In addition to the two atomic blasts, this included a series of experiments with the rather misleading name of 'minor trials'. At Maralinga the 'minor trials' evolved into some of the most dangerous scientific experiments ever undertaken in Australia. The earliest trials at Emu were relatively benign by comparison, though not as innocent as their codename, Kittens, might suggest.

The idea behind the Kittens trials was to develop the mechanism to initiate the neutrons that cause fission in an atomic bomb. The first four trials took place before Totem I, and the final one two days after. The exact location of the fifth Kittens test is unclear, but the first four were held at the K site about 6.5 kilometres from the Emu airstrip and it is likely the fifth one was near here too. The Australian Government was not notified about the trials, despite their toxicity – they used beryllium and polonium-210 (the Kittens tests held later at Maralinga used uranium-238 instead). Indeed, there was not even a formal notification mechanism. The Royal Commission into British Nuclear Tests in Australia noted this in its report, pointing out 'that it was mentioned only because the British wanted some site preparation to be carried out by the Australians'.[64] Kittens continued on much more extensively at Maralinga until 1961.

The experiments themselves tested initiators – small

devices about 5 centimetres long that set off a chain reaction in an atomic device. Inside a Blue Danube bomb, these initiators were known as 'urchins', because they looked a bit like a sea urchin. The urchin, or kitten, provided 'an intense source of alpha particles', which allowed the device to detonate.[65] These alpha particles are absorbed by the beryllium, which then emits a burst of neutrons, essential to initiating a fission chain reaction. The Kittens trials were, therefore, indispensable to the Blue Danube design and necessary to complete the requirements for an operational weapon. The Kittens tests at Emu enabled the British to refine the design of the urchins used in Blue Danube, part of the steep learning curve for Penney and his team.

The plan at Emu was to test the urchins using half-scale explosive devices without fissile material, by bringing polonium into contact with beryllium.[66] Beryllium is not radioactive, but polonium is, and during questioning at the Royal Commission, Penney conceded that some of the toxic material would have been 'spread around' in a circle of about 55 metres radius.[67] When asked why tests that spread toxic materials around in a relatively small area could not be conducted in the United Kingdom, Penney replied:

> It is very difficult in the United Kingdom to release something and be sure that it is not going to get into some dangerous thing, whereas on this range in Emu there were hundreds of square miles and no one relied on getting any water from there.[68]

This was a bold statement, given how little Penney knew about the Aboriginal use of water resources in the area. The five Kittens experiments at Emu caused significant contamination, both chemical and radioactive, although it was relatively short-lived compared to the plutonium-239 contamination at Maralinga.[69] The Kittens sites at Emu were never cleaned up.

An Australian X200 mission to Britain in February 1953 was introduced to the requirements for the Kittens trials.[70] The plan was to set up five firing stations, almost a kilometre apart. The sites required a protective sand wall so that two trailers could be sited about 365 metres upwind of each firing site in turn. The Kittens area would also have a 2.5-metre-square expendable wooden hut.[71] Eight people were to be directly involved in the trials, with support from two more for 'health and other problems'.[72] The planning for these tests noted that '[a]n area immediately around the firing sites would become contaminated and might remain so for a period'.[73] The success of the Kittens trials enabled the refinement of the Blue Danube urchin. Later at Maralinga, the British developed a different method for producing this timed burst of neutrons, dispensing with the need for polonium and beryllium.[74]

Meanwhile, at Penney's research headquarters at Aldermaston, scientists and engineers were preparing the test atomic devices for their long journey to Australia. Each device had its own team to accompany it on its flight, and they were transported separately from the plutonium fuel which was only inserted at Emu. The Totem I assembly was in the charge of G Gallie from the AWRE and Flight Lieutenant Parker, who were required to record the temperature and humidity throughout transit.[75] Likewise, the AWRE's PH Wilkens and Flight Lieutenant White were assigned to the Totem II device, including its associated packages containing a high explosive lens and other specialised items not required for Totem I.[76] The atomic devices were carried by Hastings aircraft UAH 081, which departed from RAF airbase Lyneham in Wiltshire on

21 September 1953 'carrying explosive secret freight for Emu'.[77] Once in Australia, the aircraft stopped in Darwin and then at Mallala RAAF base in South Australia. It touched down at Emu on 29 September to offload its cargo, before flying on to Woomera. At Mallala, the aircraft had an overnight armed guard.

There were two rehearsals for Totem II before its detonation, when a concrete replica of the device was hoisted up the tower, to check the first circuit equipment. This was not done for Totem I, indicating the greater care taken with the Totem II device.[78] Those in charge of the preparations were warned not to allow any dust onto the high explosive components of Totem II during the fiddlier set-up arrangements that involved twice removing these components in the dusty Emu environment.[79]

During the 1950s Britain also formulated plans to create an early form of intermediate-range ballistic missile to carry nuclear warheads. Ultimately these came to nothing. Blue Danube was a device of large dimensions, which ruled it out from being attached to a missile. The only time it was dropped from an aircraft was as part of Operation Buffalo at Maralinga in 1956. After the V bomber fleet became obsolete in the mid-1960s, the country's nuclear deterrent became the submarine-based Polaris and later Trident missiles.

The military applied pressure for the rapid development of atomic weapons, Penney said, which had the effect of bringing the civilian side of atomic energy along behind it:

> The military side gave tremendous impetus and urgency to the program. It made it possible to get things done more quickly than would have been possible otherwise. We always regarded the Calder Hall reactor as dual purpose –

to produce plutonium for use in weapons and to generate electricity. There is no phase of the program that is completely military. We looked in two directions from the start. The thing that surprised us was the civil side came on so quickly. The military program brought priorities and schedules and carried the program with it.[80]

Britain was the first country to produce atomic energy for civilian use.

Lindemann's vision for post-war Britain was a country in charge of its destiny and powered by nuclear energy, both physically and economically. He did not want the nuclear enterprise bogged down in the civil service, either. He believed a profit-making venture could help revitalise the country after the war had sapped its energy. Lindemann worked relentlessly for British atomic research to be removed from the Ministry of Supply and placed into a non-departmental organisation. Churchill's son-in-law, Duncan Sandys, as minister for Supply, stood to lose his grip on the British atomic program, so when Lindemann pursued the proposal at a top-secret Cabinet meeting, Sandys raised a variety of complications. These included, among other issues, the fact that the government would have to prop up a corporation since atomic energy was years away from being profitable; that the salary demands of hard-to-find scientific and engineering staff could not be contained, making research extremely expensive; and that there was no security benefit of a new organisation.[81]

Lindemann countered 'that the great disadvantage of the present organisation for the control of atomic energy was that there was no proper direction at the top'.[82] Industrial use of atomic energy was still some way ahead, he said, but the country was already producing fissionable material as part of a large industrial undertaking. He wanted the vigour of a commercial organisation brought to bear on the challenging problems of the nuclear age and a board of directors and a managing director to run the organisation while acknowledging some government involvement, given the security issues at stake.[83] Lindemann was frustrated by the delays to plutonium production. Getting new reactors up and running was slow and laborious, and Lindemann blamed the bureaucracy. He said that recruiting the best people was hampered by low civil service salaries, particularly production engineers, since their skills were rare and in high demand. Even nuclear scientists were hard to retain. Lindemann pointed to the staff turnover in William Penney's HER department of around 10 per cent each year.[84]

> It was not only that the Ministry of Supply was losing good men whom it could ill spare; there was also the danger of letting men go into outside employment in possession of all the secret information which they had acquired while working for the Ministry.[85]

In a private organisation, salaries could be kept secret and individually negotiated, allowing for more flexibility in appointing the best people. Lindemann argued that the nuclear enterprise required a new way of thinking not bound by the maddening slowness of the civil service. According to historian Margaret Gowing, Lindemann 'believed atomic energy was uniquely different from all other Government scientific projects, so much so that it justified quite exceptional treatment'.[86]

Lindemann further argued that American distrust of the British civil service 'could never be satisfactory under Civil Service conditions. Under a corporation, all contracts would contain a clause enabling them to be terminated, if necessary, on security suspicions'.[87] Lindemann's plan for an autonomous organisation was realised in 1954 with the establishment of the UK Atomic Energy Authority, responsible for both civilian and military uses of atomic energy.

By the conclusion of Operation Totem in 1953, and despite the McMahon Act, the Americans were more relaxed about Britain pursuing atomic technology, and were reactivating some backchannels with the United Kingdom. The McMahon Act's days were numbered. Lindemann let Churchill know the good news in a note sent on 12 November 1953:

> You will be glad to learn that the Americans have now definitely agreed to exchange information with us (and the Canadians) on effects of atomic bomb explosions. This will cover the whole field of weapons effects, troops, tanks, planes etc. Having carried out some scores of tests they naturally know a great deal more about this than we do.[88]

The new arrangements were to be put into law by an amendment to the *modus vivendi* that had governed the so-called 'special relationship' between Britain and America since the late 1940s. They were the first extension of Anglo–US collaboration since the McMahon Act came into force in 1946. 'I am glad that there should be at least this one definite positive result of our recent negotiations in Washington. The intelligence matters which we discussed there are also, it seems, going well.'[89]

Only a year separated the heady success of Operation Hurricane in October 1952 and the commissioning, development and use of a totally new desert test site in the Australian wilderness. A year to prepare two highly experimental 'austerity' weapons and ship them halfway around the world for testing. If Hurricane had been a huge success for Penney's elegant Blue Danube design, a win for the atomic weapons purists, Totem would test a much more pragmatic concept: whether a broke nation with few resources could create enough atomic bombs on the cheap to keep the Soviet Union at bay and win back the United States. In pursuing this goal, Lindemann was a powerhouse, without whom the onward momentum of the British atomic project might have been slowed, if not halted. Through his advocacy, and his closeness to the British prime minister, the scene was set for sound and fury in the Australian desert.

Chapter 4
SOUND AND FURY AT EMU

> The 'atomic weapon' detonated at Emu Field is now officially described as having a 'relatively small fissile content' ... How much below the dimensions and power of America's so-called 'nominal' atomic bomb it is now possible to reduce a practicable 'atomic weapon', the mere layman has no means of knowing. But the fact that 'Operation X200' produced a flash infinitely brighter than the light of the morning sun, and a shattering noise audible hundreds of miles away, is proof enough that even a comparatively small amount of fissile material is of the deadliest potency.
>
> 'A Flash and a Roar', Adelaide *Advertiser*.

Summer arrived early at the X200 site in 1953, causing problems with the meteorological forecasts. Unseasonal rain and a reduction in the frequency of suitable wind conditions led to a one-week delay in the Totem I firing.[1] In fact the conditions should have put the brakes on the whole series. That didn't happen. The British nuclear tests in Australia were riddled with recklessness around safety. The unseasonal rain was highly significant because it created the conditions for the black mist that rolled over settlements to the north-east of Emu Field.

The Totem series put two different kinds of low-yield

plutonium weapons under the AWRE microscope and is often described as a 'comparative' series because it was designed to show changes in performance between devices powered by differing fuels. Totem I was a Blue Danube design, similar to the one tested at Hurricane, but modified to fit the new V bombers that would carry it. Totem II was more experimental, also based on Blue Danube, but its differences remain secret. The two 30-metre towers built about 1.8 kilometres apart to hold the test devices were strong enough to hoist weights of up to 4.5 tonnes, one of the Blue Danube design parameters.[2] The towers and the tests held on them were codenamed 'Tom' and 'Cat', or Tom-Cat.[3] Tom-Cat had been the project's working codename prior to Totem, and the name lived on at the test site.

The British fretted over exactly what to tell the Australians and finally settled on a form of words sufficiently vague while sounding authoritative. The advice as first written emphasised the strategic reason for the tests: 'In order to determine the minimum size of effective weapons that can be made, it is necessary to carry out as a matter of urgency a comparative trial involving two weapons'. However, in a secret meeting at the UK Ministry of Defence, a handwritten change to the beginning of the sentence, sought to remove the reference to size: 'In order to determine certain design information on atomic weapons'.[4]

This suggested change was quickly overruled, because yet another version was sent to the Australian Government the day of the Defence meeting. The Totem Aide Memoire of 13 December 1952 simply stated that the tests had 'become necessary in order not to delay the rapid development of the United Kingdom Atomic Energy Programme. It is intended to explode two bombs, each considerably less powerful than that detonated at Monte Bello'.

The British were trying to make small but effective atomic

bombs on a shoestring. Although wary of telling the Australians too much, they wanted them to know that deliberately small bombs would be tested at Emu. As was noted in the minutes of a November 1952 meeting of HER personnel, 'Dr. Penney outlined the need for an urgent trial to determine the smallest bomb it was possible to make'.[5] Lindemann wrote to the UK Ministry of Supply in February 1953: 'I would suggest that we tell the Australians that we are deliberately experimenting to find out how small it is possible to make weapons while obtaining a reasonable efficiency'.[6] At the same time, the British did not want their attempt to make a small weapon to be taken to mean that the Totem bombs were a step down from the larger Hurricane device. More pertinently, there was no reference to the fuel that powered the tests or the fact that this was at the core of the Totem experiments. The official line on the tests was that they would 'determine certain characteristics of fissile material'.[7]

Totem I, the first mainland test originally scheduled for 7 October 1953, was delayed until 15 October. The highly experimental nature of the bombs is underlined by the fact that the bomb-makers made poor predictions, and got the yields of both bombs significantly wrong. They even wrongly predicted that Totem II would be twice the explosive power of Totem I.[8]

Before Totem, the British knew almost nothing about fallout from atomic weapons. They did little in the way of fallout studies from Hurricane and proceeded quickly to Totem without understanding the phenomenon – certainly not in relation to the particular geography and meteorology of central Australia. To what extent was this a 'known unknown' or an 'unknown unknown'? The question matters, because it speaks to the issue of whether the British knowingly experimented on human populations within range of the Totem bombs by allowing radioactive fallout to spread across the land. Were they callous, even murderous, or just ignorant? British meteorologist Edward Siddons, giving evidence at the Royal Commission said:

> The Totem explosions were the first occasion from which the Atomic Weapons Research Establishment obtained any real data on fallout; very little was obtained from Hurricane, for example. Of course the immediate reaction of the scientists given that data was to check it out with what was predicted and to attempt to improve the methods of prediction. That was done during the two years or so after Totem and before the next Australian trials.[9]

Those later analyses were of no use to people in harm's way during Totem.

At the Royal Commission, Penney was asked if he harboured concerns in the planning stage that 'the fallout from those [Totem] tests would necessarily pass across a significant part of Continental Australia?' Penney's answer was hardly reassuring: 'They would pass around the world'.[10]

Fallout from Totem I extended to the north-east of ground zero in a narrow, cigar-shaped plume, and eventually made its way right across the country to the eastern seaboard.[11] Totem II had a different fallout pattern: its cloud was subjected to considerable windshear (sudden variations in the direction and movement of wind), being forced south-south-west of ground zero for about 2 kilometres, then broadening and veering to the south-east.[12] The radioactive clouds of both explosions crossed the Australian coast at North Queensland.[13] Doubtless, residents of Townsville and surrounding areas at the time were oblivious to the radioactivity that reached them from tests in South Australia and may be surprised to hear it now. The Woomera range would have seemed so far away from tropical North Queensland.

At the time of Totem, the British not only knew little about fallout; they were significantly lacking in meteorological knowledge. Totem I was carried out in conditions where there was little or no windshear, and consequently the atomic cloud kept its shape for longer than the Totem II cloud, and was visible as a coherent mushroom shape for up to 24 hours.[14] The remnants of the mushroom were seen by civilians on an aeroplane over Oodnadatta, over 450 kilometres to the east of Emu.[15] This was something Penney and co. were keen not to repeat and, as noted by Arnold and Smith, they agreed to explain cloud sightings away as 'rain clouds'.[16] Counsel to the Royal Commission Peter McClellan put this question to Penney: 'As a consequence the fallout remained concentrated in that cloud for a longer period of time?' Penney: 'In the smaller area of space, correct, yes, that is right.'[17] In the case of Totem II, the wind shift at low altitude caused fallout to drop all over the site road system and other services.[18] The shock wave from Totem II also produced much more dust at the Emu site than Totem I, another marker of the quite different meteorological conditions for both tests.[19] As the Maralinga Rehabilitation Technical Advisory Committee (MARTAC) report of 2002 drily commented, '[n]either of the Totem trials was carried out in anything approaching ideal meteorological conditions'.[20] (MARTAC was an Australian Government organisation formed in 1993 to oversee rehabilitation of British atomic test sites in Australia.)

Planning for the Totem tests did not always go smoothly. Penney and his team were flying blind in many ways, including the logistics of atomic weapons testing. The maritime nature of Operation Hurricane in October 1953 was unique. Now the Brits had to become familiar with terrestrial testing in harsh and arid conditions, made more difficult by an acute shortage of funds. Penney's original lavish plans for the test, which included the use of helicopters, had to be scaled back. Head of the UK TOTEX committee, Air Marshal Thomas Elmhirst, was informed by Penney about the inability to use helicopters and quoted him in a note to FC How at Supply saying: 'In view of the difficulty which is arising in our request for helicopters, and because they are not absolutely essential this time, I feel that we might withdraw our request'.[21] Elmhirst was personally embarrassed by this development, writing to the controller of atomic energy, Sir Frederick Morgan (who had succeeded Portal in 1951): 'I am most distressed that I misinformed you on this question but I had very definitely got the idea, during five days at Fort Halstead, that these helicopters were essential. It now appears that, though desirable, they are not absolutely essential for Totem'.[22] No helicopters flew that October at X200.

Penney and the military wanted different things from the atomic tests, and there were many arguments and compromises along the way. Penney was interested in weapon *performance*, while the Joint Chiefs of Staff were interested in *effects*. The British Army wanted to discover 'what would happen if any atomic bomb were dropped on, for example, a division concentrated on a river and about to cross it, a corps area with depots or a port of embarkation'.[23] In particular, they wanted to know the effects of the explosion on equipment, such as bridging, ammunition, food supplies and vehicles, and approached the Australian military to discuss placing equipment into the blast zone of the test.

One of the experiments required by the military was to see

what would happen to a battle tank placed in the midst of an atomic explosion. The Australian Army obliged, and a crew led by Captain JG Monaghan left Puckapunyal on 17 July 1953 to deliver the massive vehicle.[24] The Centurion Mk III tank, registration number 169041, was situated in the Totem I blast zone. The Australian Army fitted special filters to the tank. The vehicle had about a quarter of a tank of fuel and normal oil levels, to see the effect on inflammable materials. The engine was kept running throughout the test, and was still running after the detonation. Dummies representing crew were provided with film badges to test for radiation.[25] Film badges were plastic holders containing a piece of film similar to dental x-ray film. The film could be exposed in the presence of radiation and later developed like photographic film to determine the radiation dose received by the wearer. An analysis of the film badges showed that living crew would have received a dose of radiation sufficient to kill them within a few days.[26] After the Totem I explosion, the Centurion was driven from the explosion site under its own power. A later review of veterans' entitlements found that the 'crew members who travelled with [the Centurion tank] on its journey from Emu Field to Puckapunyal were exposed to radiation. No crew member wore film badges for any part of the journey'.[27] This tank was later put into service in the Vietnam War and was celebrated as the 'atomic tank'. An AWRE report produced in 1954 noted rather bluntly that: 'It appears probable that tanks are more resistant than previously believed, but the crew is easily the weakest link in the equipment. The soundness of design and robustness of construction of the Centurion is generally confirmed'.[28]

In addition, six Australian-built Mustang aircraft provided by the RAAF were exposed to the Totem I blast, at a variety of distances from ground zero, and at various angles to the blast.[29]

This expensive sacrificial gear was joined by a ship's funnel, concrete shelters and various kinds of measuring equipment to capture aspects of the blast.[30] These experiments, to determine how military equipment would behave in a nuclear attack, were of greatest interest to the British military but of virtually no interest to Penney, who could have done without them. In general, Penney was irritated by military demands. His focus was on the scientific and engineering aspects of his bomb tests: he wanted to see exactly what it would do when detonated.

Four camera towers were constructed on the site. Since they were found to be too unstable in the face of likely blast waves from the tests, they had to be secured with guy ropes. The photographic team was vexed by the organisational side. Head of the AWRE photographic observation team, EW Walker, noted in his top-secret report after Totem, 'the task of record photographers would be easier if they knew in advance what records were required and when they could take them. Improvement in this respect is expected to remain a pious hope'.[31] Walker deemed the photographic equipment used at Totem 'quite inadequate'.[32] The photos captured of the two tests were, however, dramatic and evocative, even if technically they were not entirely satisfactory.

The small number of official on-site personnel included the Australian observers Ernest Titterton, Leslie Martin and Alan Butement. This trio joined a range of key AWRE personnel led by Penney, including Noah Pearce, in witnessing Totem I from the lookout. Australian personnel were also listed for duties in radiochemistry, photographic observations, army equipment, meteorological services and communications.[33] British scientists, who started arriving in batches of ten from August, were allowed some relief from their new surroundings at Emu:

> In view of the very irksome conditions under which these scientists will be working for a long period, Head of [the UK Ministry of Supply staff] is anxious that they should be given facilities for some recreation during their stay in Adelaide. Accordingly he has requested us to provide one or two cars so that the men can visit the inner S.A. countryside during their two days spell in Adelaide.[34]

No doubt the excellent wines from the area would have been among the compensations.

Len Beadell's work in setting up the Emu site was highly praised by Major-General Jack Stevens, head of the Australian Totem Panel. Stevens wrote to Elmhirst saying:

> All instrument and test piece sites have been positioned by Beadell – to an accuracy of one inch ... All main recording sites and towers to six inches ... Strongly recommend that nobody [wastes] time trying to improve on Beadells [sic] figures. Because of accommodation difficulties better for project if anyone with time to spare for amateur surveying be smartly moved out of camp.[35]

Beadell was a favourite with everyone associated with the Totem tests for the precision of his work, and also his humorous demeanour, in keeping with the frontier atmosphere of the site and embodying the Antipodean style that the British found endlessly entertaining. Well after the bomb program in Australia ended, Beadell claimed to have been present for the Totem I blast. Writer Ian Bayly disputes that claim, backing this assertion with Beadell's own evidence at the Royal Commission when he clearly stated that he 'had missed the first one'.[36] Still, Beadell dined out for years on his tales of being there, including in a famous talk at the Shepparton Rotary Convention in 1991. He may well have been present for Totem II on 27 October. If Beadell wasn't present for either of the Totem bombs, he was a particularly gifted and imaginative storyteller who maintained the story with considerable detail in his final career as a raconteur.

The limitations of the camp became apparent as the date for Totem I approached, and 'tourists' – people who were more curious than essential – became a problem. A cable from William Cook at the Ministry of Defence to Brigadier Leonard Lucas, who was in charge of construction at Emu, took a hard line:

> Neither water nor accommodation will carry more than 300 which figure given by U.K. Feb and reiterated constantly since. If hordes of boffins and tourists insist on coming then compensating number of construction force moves out and boffins do own laboring. Are all intending residents aware no repeat no batmen no repeat no mess stewards and damn all water other than salt.[37]

Senior officers would have to do without the help they usually enjoyed, and ring-ins would push out someone essential.

While Operation Hurricane had included a range of peripheral tests, the trials at Emu were of necessity going to be more restricted. Most equipment had to be brought in by air, placing immediate limitations; Penney estimated that he would need 'between 60 and 70 tons of crated scientific instruments'.[38] A meeting between Penney and representatives of the military services, as well as Civil Defence, set out the various wish lists. Penney said no to most of the ideas put forward.

The British Army was particularly interested in fallout data. Penney countered that 'very accurate data had been obtained at the Monte Bello trials on fall-out up to 6 or 7 miles down wind. The forthcoming trials would include a very detailed survey of fall-out up to much greater distances, in view of the obligations to Australia'.[39] In fact the British learned very little about fallout from Hurricane, so Penney's comments are inaccurate. They learned much more about fallout from Totem.

The small, dedicated team at Emu worked through the night to be ready for the Totem shots, both of which were detonated at 7 am. The site was lit up with floodlights as each designated team fulfilled its role in the preparations. It must have looked like the set of a science fiction movie before the dawn light appeared at the distant horizon (sunrise was 5.34 am on 15 October). Totem I was fired on Penney's orders; he directed the display from a rise on the plain known as the lookout. This is the only high spot on the site and affords a clear view right across to the two ground zeroes 6 kilometres away. The sight of the atomic cloud lifting from the mulga scrub was awe-inspiring. Multiple cameras captured the conflagration and revealed a terrifying fireball and rushing shock wave.[40] The atomic cloud rose to a height of 4572 metres, with a radius after three minutes of 1067 metres.[41] With binoculars fixed on the site, Penney and the small group of witnesses had a prime position to see the atomic fireball turn the sand to dark glass. The trinitite created by the test contains trapped radioactive materials including plutonium.[42] Emu trinitite is darker than that at Maralinga because the soil contains more iron. As well as scattered, randomly shaped glass fragments, there are also 'beads': teardrop-shaped, solidified material created when the soil and limestone beneath the bomb blast liquefied. The beads are found only at the Totem II ground zero, indicating the slightly different terrain even though the two sites are close.[43]

Adding to the drama of the occasion, 30 rockets were fired into each Totem explosion, at one-second intervals, from 12 seconds after detonation. The rockets created the characteristic smoke trail lines shown in many pictures of atomic tests. Each rocket was fitted with a quartz fibre filter for the collection of radiochemical samples. They passed through the cloud at 1829 metres, then fell into an open area for collection and later analysis in laboratories at Salisbury.[44]

A British military participant kept a handwritten diary of Operation Totem that provides a colourful account of events as they unfolded. His name is not shown on the original document but he was in the Mechanical Effects Group led by AC Purdie from the AWRE. The entry from 15 October 1953, the date of Totem I, begins, 'Weapon fired according to plan. Stopwatch observations made at 3 by ME [Mechanical Effects] Group and Schofield. Encouraged by estimate of 9 [kilotons] of TNT but later dismayed because had misread graph. The answer appears to be 17KT – surely too big! Must check later'.[45] He was right the first time: Totem I yielded just over 9 kilotons.

The diarist recorded the frenzy of activity after the blast, including retrieving gauges (many of which were completely destroyed), weighing tubes and taking measurements, and fitting in a late snack lunch at 2.30 pm. 'D.AWRE [director, AWRE, William Penney] has been pressing for results almost from the word "go". He continues to do so until about 5 pm'.[46] It was a busy, eventful day for everyone. 'In evening we relaxed. General atmosphere of jubilation'.[47]

Tony Brinkley, Ernest Titterton's laboratory assistant, who worked on Titterton's Emu experiments, gives another account. He describes getting up in darkness and going to the elevated lookout position with a small, elite group that included Penney and Titterton to witness the shot:

At zero instant I saw a flash of light through my eyelids and felt a surprisingly hot lick of heat on my neck. I turned around with everyone else and watched the fireball. There was no sound but we could see a disturbance racing across the desert. The sound hit with an incredible instantaneous crack, like a rifle shot, not a big gun, even though old hands like Titterton and Penney ducked. There was no boom or roiling echo such as the press and VIPs heard twenty miles back at the camp site. I remember being greatly disappointed at the bang.[48]

Brinkley was required to go close to ground zero only 90 minutes after firing:

> I was dressed in a long underwear garment, thick wool socks, a boiler suit type of protective clothing, helmet, respirator, gloves and rubber boots. I was accompanied by a young man who was to monitor radiation levels while I did my work. I drove the jeep out along what was left of the track and began my work. Eventually I went in as far as zero position where the sand had been fused to coarse glass by the heat. I was probably at the glass for about a minute; my radiation monitor got so worked up that I sent him back to the jeep. We had a bit of difficulty getting back because everything looked completely different and I couldn't find the track. Also I got the jeep tangled with a lot of control wires. I gave up trying to free them and just drove off towing the lot. After working about an hour all told we got back to the monitoring area at ~11 a.m. My monitor went one way, and I another. I was found to be heavily beta contaminated on the outside of my protective clothing. At this stage I hadn't removed my respirator or helmet. I can't be sure, but I think

was hosed off before I removed my clothes. When I took my respirator off half a cup of water poured out of it; the same thing with my gloves and rubber boots. Then with no clothes on at all I was monitored on the skin and put under a shower for about 10 minutes, then monitored again. That afternoon I was flown from X200 site to Adelaide and next day I flew to Canberra.[49]

It can't have been reassuring for Brinkley when Titterton told him soon after that his film badge was completely black and therefore his radiation dose could not be assessed.[50]

An Australian party of senior government officials and military officers inspected the Totem I site five days after the test. The group included WR Blunden and Lieutenant WF Caplehorn, who remarked on the destruction in a report they prepared:

> The ground for about 150 yards in radius about the tower was seared, and, near the centre appeared to have been glazed by the heat of the flash ... The mulga scrub had been completely obliterated over about 300 yards radius from the point of the burst, shattered for a further 400 to 500 yards, broken down or knocked about for a further 500 yards.[51]

The group carried radiation detection equipment:

> All instruments had exceeded full scale reading. It was estimated that the party had received one roentgen by driving around the centre of the contaminated area in Land Rovers. It was of course a perfectly safe dose, but shows the danger of staying in the area of fall-out even several days after the explosion. Several hours spent in the area would

have resulted in a dose greater than the maximum safe figure.[52]

The Blunden and Caplehorn report indicated that only a limited amount of equipment was exposed to the blast and 'it seemed to us that more use could have been made of this opportunity without overtaxing the resources of transport and accommodation at X200'.[53] While Penney managed to keep most of the focus on the novel Totem weapons, some military tests were given the go-ahead. Slit trench measurements, for example, were designed to measure how radiation permeated 'typical weapon-pits' that might be used in actual warfare, and to see if a horizontal slab of earth placed above the pit provided overhead cover.[54] British military leaders believed that atomic war was imminent and wanted to protect their military assets from an atomic weapon. Totem was a perfect opportunity to work out what changes they should make to their operations.

The slit trench experiment followed on from similar activities during Operation Hurricane. The work involved creating 'foxholes' of the kind that military personnel could conceivably use during wartime. They were just over 1 metre in diameter and just under 2 metres deep, and were dug in pairs at each of five sites, with two film badges placed in each. Unfortunately, the crew tasked with this job ran out of film badges. Consequently, E Drake Seager from the UK War Office for Operation Totem reported, 'the precision of the observations is in many cases doubtful'.[55]

Drake Seager's team also carried out tests to determine exactly how hot it got close to an atomic explosion. Thin strips of wood, known as laths, were coated in Detector Gas Paint No. 1 or No. 2 and placed in the slit trenches. This paint turns from green to red when subjected to heat of more than

1 calorie of incident heat per square centimetre, roughly the amount of heat generated by a one-second burst of a cigarette lighter. The aim was to determine how far down the wall the heat from an atomic fireball would travel, potentially harming any foxhole occupants.[56] The tests showed that the total heat dose from the thermal radiation reached the dangerous level of over 2 calories per square centimetre close to the top of the foxhole. People sheltering in foxholes during a nuclear attack would not necessarily burn to death, therefore, unless they poked their heads up too far. Another slit trench experiment involved placing 'toothpaste tube' blast gauges in the bottom of the foxholes to record the blast wave at 1.83 metres down. These aluminium tubes shaped like toothpaste containers would collapse in response to the pressure of the blast and enable the peak blast pressure to be measured.[57]

Drake Seager's team had a dark sense of humour. In one experiment they placed gas masks on wooden heads coated with heat detecting paint. 'As an additional sartorial touch CD berets were put on the heads at the appropriate angle.'[58] The heads were placed on 2.3-metre-high scaffold poles at various distances from ground zero. The total heat dose was found to be up to 140 calories per square centimetre for those gas masks closest to the blast, easily enough to incinerate a person.

An array of measuring techniques was used to better understand exactly what happens upon detonation, including the heat of the explosion. Operation Hurricane had not been subjected to heat measurements, so these were new data for the British. In all, they used four methods to measure various physical parameters: photo-electric, photographic, bolometric (electromagnetic radiation) and calorimetric (heat).[59] All the measurements showed that Totem I was a more intense blast than Totem II, confirming that Totem I had a higher yield:

9.1 kilotons compared with 7.1 kilotons. The heat from both was mind-boggling: 6726 degrees Celsius and 6226 degrees Celsius respectively.[60] Ernest Titterton was involved in optimising the equipment to make these measurements. He was also involved in measuring the output of fast and slow neutrons, vital data about the physical effects and yield of an atomic explosion.

Titterton's ambiguous status as an insider–observer gave him various advantages – trusted (at least until the early 1960s) by the Australian authorities, and used by the British as evidence of Australian inclusion in their tests. His nuclear elite status made it possible for him to carry out experiments as part of Operation Totem, the only Australian-affiliated scientist to do so. Specifically, his experiments on fast neutron measurements allowed him to calculate highly secret information about the explosive yield of each of the Totem devices. The head of the Australian Totem Panel, Jack Stevens, gave him explicit permission to carry out these experiments, after he had written to request it.[61] Titterton got three junior scientists from ANU to help him: Tony Brinkley, John Carver and AC Riviere. Earlier, Elmhirst had written to let Stevens know that Titterton's research was welcome at Totem:

> We should like details of proposed experiments, number of staff and details of any special equipment not available from Australian resources ... Project to be graded top secret and we presume that your security authorities would be fully confident and satisfy themselves first that any staff working on project are completely cleared for security.[62]

John Carver many years later became head of the ANU's Research School of Physical Sciences and Engineering. At the time of Totem, he was 27 and not long returned from the Cavendish Laboratory at Cambridge with a PhD in physics.[63] Together with DG Vallis from the AWRE, Carver and Titterton carried out measurements for both Totem I and Totem II. In a note to Titterton discussing some of his technical findings, Carver wrote, '[i]t has been fun to see one of these shows from the inside and to receive so many "signals"!'[64] There were data galore for Titterton and his young team.

The Titterton experiments involved placing plastic envelopes containing sulphur, selenium or nickel into shallow holes at regular intervals ranging from 92 to 914 metres from ground zero, then examining the radioactivity on the plates after the blasts. The preliminary report from Totem I indicated that all samples were recovered except for those placed 92 metres from the blast. They found that radioactivity was high on the close samples, and those at 137 metres were fused. Titterton indicated that more complete results would be available before Totem II.[65]

Fission weapons are based upon the way neutrons behave. Atomic nuclei are wrenched apart by neutrons fired at the fissile material, releasing the energy that creates the blast. Titterton sought to understand what both fast and slow neutrons were doing as they escaped from the device. According to Titterton's report, the fast neutrons escaping from a fission bomb carry 0.03 per cent of the total energy and can travel a long way through the air upon detonation of the weapon: 'They can penetrate considerable distances through air and, because of their great effectiveness in causing physiological damage, e.g. cataracts, they represent a serious hazard out to a considerable range'.[66] The experiments were intended to estimate the hazard in relation to distance from the blast and increase knowledge about the damage done at different distances from an atomic explosion. Only a practical experiment could do this, said Titterton; it could not be accurately calculated.[67]

Some of his samples were accessed at 30 hours after Totem I, and the rest 48 hours later, except for the 92-metre samples, which could not be found, presumably destroyed by the force of the blast. After Totem II, the closest samples were also missing, as was the 365-metre sample. The recovered samples were flown to Salisbury for analysis. Titterton's experiments found that the lethal range for both weapons tested at Emu would be 'a little over 1000 feet'.[68] Anyone within that range – 305 metres – would die immediately.

Titterton and his team also measured slow neutrons; these are less dangerous to living creatures than fast neutrons but released in far greater numbers.[69] This experiment involved burying photographic plates in cylinders, which they experienced with some difficulty, owing to the thin soil. The plates exposed to the Totem I explosion were flown to Canberra for processing. Because the explosive yield of Totem I was twice as great as had been predicted, many of the plates were unreadable. In his report, Titterton said, '[a]fter the bad experience with T-1 the layout of plates for T-2 was changed'.[70] But again, the yield was underestimated, and the much larger yield meant that usable plates were obtained from only three locations – 640, 731 and 1097 metres. In the end, the slow neutron experiment showed that the photographic plate method was suitable only for tests of weapons with a yield of under 1 kiloton, well below the yield of the Totem tests.[71] Titterton had access to American data to relate to his results, but he found that he was comparing apples with oranges. His results for both fast and slow neutrons indicated much smaller dose rates than shown in the American data, casting some doubts upon his own measurements.

Titterton was also responsible for detecting radioactivity at remote locations, including the north coast of Queensland. Australian military personnel under the command of JM Crofts in Townsville manned and checked seven instruments regularly both before and after the blasts.[72] The personnel were instructed to contact Titterton immediately if they detected any radioactivity. They also measured radioactivity in locations where rain had fallen within 80 kilometres of a measuring station, since rain can carry fallout.[73] Crofts' report indicated radioactivity was detected at Bowen, south of Townsville and about 2500 kilometres from Emu Field.[74] This was the only positive result detected by the north-east coast team, though later calculations from detectors on board aircraft showing that the atomic cloud from both blasts had crossed the North Queensland coast near Townsville.[75]

Penney, reflecting later on Totem I in a letter to Jack Stevens, claimed that the week's delay for Totem I was worthwhile 'because the radioactive cloud passed over Australia without causing the slightest anxiety'.[76] The two governments and their media liaison teams had done all they could to dismiss any risks, and avert public anxiety.

The Australian Commonwealth Meteorological Service set up a meteorological office at Emu and provided communications, equipment and staff towards the meteorological effort. Two Australian forecasters were assigned to the post, and three assistants.[77] According to a top-secret 1954 report on meteorological aspects of Totem, 'the Australian staff were anxious to carry out the purely meteorological side of the work without U.K. assistance. In consequence all plotting and analyzing of weather maps, preparation of forecasts, and taking and recording observations was carried out by them'.[78] Despite the effort, the meteorology at the site was not up to the task of managing the complex process of predicting and tracking atomic clouds. Both Totem clouds went rogue once they had been detonated.

Winston Churchill sent a cable to Robert Menzies once Totem I had been exploded. He was quoted in a press statement:

> I have learnt with keen interest that the first of this year's atomic tests has been carried through successfully. The assistance of the Australian Government has been invaluable and success could not have been achieved without the closest collaboration between our experts. Thank you so much for your help.[79]

No doubt Churchill and Lindemann were pleased with the news from Australia. The prime minister's mood was probably further lifted when he received the Nobel Prize in Literature just days before Totem II.[80]

Totem I had the biggest impact on the local population, although it has never properly been measured or even understood. Yet the British authorities have long been more concerned with the security of Totem II, which was more experimental. So long as they maintain the utmost secrecy about Operation Totem, the exact nature of Totem II will not be known outside the ranks of those with high security clearance. This is true even all these years later. Minister for Supply, Duncan Sandys, visited Australia just before Totem, and, with typical vagueness, described Totem II: 'The object of this second major explosion was to obtain certain important scientific information which we required in connection with our weapons programme'.[81] The concerns about Totem II security peek out tantalisingly from various official documents we are allowed to see. For example, AWRE scientific director

Charles Adams wrote to Titterton in December 1953: 'I am sure that you will bear in mind that the performance of Totem II is still shrouded in official secrecy'.[82] He included some photos of the Totem II shot, which, 'as you know, have not been issued to the Press and therefore retain a security grading'.[83]

Lindemann, as a scientist and insider, knew exactly what Totem II was. He was adamant that no media or politicians be given seats for the second show. In his instructions about media policy for Totem, he said: 'Press not to be nearer than some 14 miles and to be allowed for the first test but not for the second (which is an experiment which will probably fail but which we must try because of the enormous advantage of it if it comes off)'.[84] Britain did not want any witnesses to a failure. Apparently, it didn't fail, despite official concerns.

Both Lindemann and Sandys made a point of not being at Emu for either Totem I or Totem II, despite being in the country around the time of the blasts, in an attempt to keep Australian politicians away as well. The head of the Prime Minister's Department, Allen Brown, conveyed this in a note to Menzies: 'I have received this little message from [Ben] Cockram [office of the UK high commissioner to Australia] who is in Adelaide with Lord Cherwell [Lindemann]: "U.K. Minister has said he is only too anxious to set an example of being elsewhere."'[85]

The British wanted the operations at Emu to be witnessed by as few people as possible, limiting not just Australian politicians but also the Australian military. A letter from Brown to Stevens noted, '[o]ver the weekend I received a message from Cockram, who is in Adelaide with Lord Cherwell. It merely read – "The arrangements stand as regards authorizing the Australian Chiefs of Staff to go to tests with similar facilities to those given to the press."'[86] While the heads of Australia's military were going to have to slum it with the press, their counterparts from the United Kingdom would be part of the official party. The Australian party included the chief of staff of each of the three services or his representative, two in the case of the army, each with a staff officer, making a total of eight, plus a UK scientific liaison officer in Australia.[87]

Brown himself was opposed to too many people attending. He told Menzies:

> We, ourselves, have a request from one of the Opposition Members of Parliament and, of course, from various people who claim interest, such as Professor [Harry Messel]. It does seem to me, however, that if non-scientific observers are to be present there is a case for at least one or two of your own Ministers to attend if you so wish. My own personal opinion is that it is all nonsense for any of these people to be going, but I do not know whether the request from the Australian Chiefs of Staff was made with Ministerial approval or not, nor do I know whether you would wish to interfere to the extent of saying that you wish that either they or their opposite numbers from the United Kingdom should not attend.[88]

He added sardonically, 'Incidentally, Moses seems to have been added to the party, because I have heard no further mention of the shortage of water which was such a prominent feature of the discussions with the Press'. Perhaps a touch of cynicism about the hardship claims being made about Emu? Handwritten on this note, possibly by Menzies, are the following words: 'Cockram tells me Cherwell has sent severe "blast" to Sandys on subject of observers. C wants it as an operational test – working scientists only.'[89]

Like Totem I, Totem II had personnel out of bed early for a

7 am blast. The atmosphere was unstable that day, and there was a strong windshear.[90] Nevertheless, the detonation occurred on schedule with only the on-site military and scientific personnel to witness it. The unstable atmosphere allowed the cloud to rise well beyond what was expected; it just kept on going up and up. While there was some confusion in the calculations, the Royal Commission accepted that the cloud rose to about 6126 metres, although some estimates put it at 8534 metres.[91] Then the windshear broke it up, making tracking difficult. The RAAF could not locate the cloud beyond about 500 kilometres east of Emu, according to the Royal Commission.[92] Not being able to predict the cloud height and movement of a mushroom cloud is a major issue for civil aviation. With good reason some RAF crew became concerned that non-military aircraft might unknowingly expose their crew and passengers to a wandering radioactive cloud.

While the Totem II cloud started out brown, it turned white after about four or five minutes, and soon took on the classic cirrus 'anvil' shape of a thunder cloud.[93] Then, like a will-o'-the-wisp, it dispersed and was seen no more. Its radioactivity, however, travelled freely across the continent.

Churchill was aware that fallout and other indirect effects of an atomic blast could cause more problems than the initial blast, and mused about it in a Cabinet briefing document in 1954.

> I was informed by our nuclear experts in August of the developments set forth in the Minister of Defence's paper of December 9. They are now embellished by the statements of the effects of 'fall out' and the intricate consequences of windborne radio-active matter on animals, on grass and vegetables, and the passing on of these contagions to human

beings through food, thus confronting measureless numbers who have escaped all direct effects of the explosions with poisoning or starvation or both.[94]

Nevertheless, Churchill maintained that Totem I and Totem II had minimal effects on Australians and Australian territory, while acknowledging the far-reaching harms these weapons could wreak.

But Churchill was a long way away. The two atomic tests held at Emu Field sent shock waves through Anangu lands and did lasting damage. In some ways Operation Totem could be considered the most damaging of all the British tests for the Aboriginal people of South Australia. As historian Heather Goodall puts it, 'in the Maralinga story, there were more bombs and catastrophic radioactive waste and damage done to the landscape, but the Emu blasts were so experimental and there was so little known about them and it all happened so quickly'.[95] Operation Totem was an uncontrolled experiment on human populations that unleashed a particularly mysterious and dangerous phenomenon that is still being debated.

Chapter 5

THE UNKNOWABLE BLACK MIST

The old people were frightened. They reckon it was mamu. It is a pretty hard word to try and translate into English, but it something that they use a lot, and I think it is something they would not understand, so we say it is mamu, you know. It is something that could be bad spirit or evil spirit, or something that we are not sure what it is.

Yami Lester, Royal Commission hearings at Marla Bore.

The Grim Reaper had been created, and, standing tall, it lurched across the countryside towards Wallatinna and as the Reaper moved north-eastwards it spread toxic and lethal particles over everything in its path.

Ray Acaster, British meteorologist.

The scene is almost unfathomable. An esteemed Australian judge and former Labor politician, sitting in a rough garage in the dust of central Australia, sleeves rolled up, holding hearings as part of that most colonial-sounding enterprise, a

Royal Commission. But Diamond Jim McClelland was never one to stand on ceremony, especially one redolent of the faded empire that he came to understand had desecrated parts of his country. His nickname had been bestowed because he favoured the finer things in life, but he was not one to tug the forelock.[1] So he took the Royal Commission into British Nuclear Tests in Australia to the outback, to Marla Bore and Wallatinna, among other locations, in the course of his enquiry. As scholar Kingsley Palmer put it:

> The Aboriginal component in the Royal Commission was important because it was meant to symbolise Australia's concern for her indigenous people ... It allowed the government to indicate through the vehicle of the Commission that the act of testing bombs in Australia was a British decision, supported by those with Anglophile tendencies with general (but predictable) colonial attitudes to those Aborigines who might be inconvenienced.[2]

McClelland scorched the United Kingdom in his findings, in part because of the courageous testimony of the Aboriginal people who suffered the effects of being British guinea pigs. The Royal Commission made strong statements about the harm caused to Indigenous people by the British bombs. McClelland was moved by what he heard at the outback hearings, and it affected many of his conclusions and recommendations.

Marla Bore is on the Stuart Highway and close to the railway that bisects the country from east to west. Wallatinna is about 20 kilometres to the west. At the time of Operation Totem, Wallatinna station was owned by a white man, Tom Cullinan. Cullinan employed only Aboriginal people, and about 34 lived at the homestead in 1953.[3] Walter MacDougall, the Commonwealth Government's NPO, reported that he knew of 172 living in the vicinity of Wallatinna at that time.[4] Diamond Jim presided over Royal Commission hearings at Marla Bore on 20 April 1985 and at Wallatinna Homestead the next day (by coincidence, the actual birthday of Queen Elizabeth II, whose Coronation was celebrated so enthusiastically by troops at Emu in 1953).

As he was about to hear evidence at Marla Bore, Justice McClelland voiced what many were thinking – that by coming to central Australia the Royal Commission gave those affected by the Emu blasts 'an opportunity ... to unburden themselves of matters which have been bottled up, quite unhealthily probably, for many years'.[5] McClelland suspended some of the more onerous rules of evidence and courtroom etiquette and made the proceedings more informal than usual. He allowed those being questioned to dwell on their reflections perhaps longer than might be the case in a bricks-and-mortar courtroom.

The Totem devices, especially Totem II, were highly experimental. Ongoing British secrecy means that 70 years on we still do not know precisely what was in the nuclear fuels that powered them. All we know from official sources is that the British were attempting to find innovative ways to save on costs and use materials they had to hand. Totem was designed to test cheaper weapons that could be mass produced faster, using higher than usual proportions of plutonium-240.

One indisputable fact is that Totem I created a mystery that will linger for as long as the British authorities choose to suppress information. Directly after the Totem I blast, a sinister dark rolling fog headed straight for human settlements at Marla Bore, Mintabie, Opal Field and Wallatinna and Granite Downs stations. Testimony decades later at the Royal

Commission by survivors is chilling. In all, 48 Indigenous people gave evidence to the Royal Commission about many aspects of British atomic testing, at various locations. Some spoke in English, some in language, translated by interpreters. The Wallatinna and Marla Bore witnesses all spoke of something extraordinary and terrible that had happened in October 1953. Their accounts of the black mist, or *puyu* (smoke/mist), were consistent, and the royal commissioners were shocked by the stories. Anthropologist Annette Hamilton, giving evidence to the Royal Commission, pointed out:

> that all these people would have had an intimate understanding of normal physical events in their environment, and all noted that the *puyu* was different to anything they had previously experienced … A dust-storm might be the most comparable phenomenon, whereas all those witnesses who mentioned wind described it at the time as a 'breeze'. Mists, which do occur rarely during the cold seasons, might be similar in some respects but they do not usually move and are white in colour.[6]

Whether the black mist was directly related to the nuclear fuel used, or whether some other factor or combination of factors sent the sinister cloud racing towards Aboriginal populations nearby may never be known.

Several attempts have been made in recent times to fathom what actually happened. At the time of Totem I, and when media stories started emerging in the 1980s, the idea of a black mist was commonly dismissed as folklore and mythology mixed with anxiety over incompletely understood events.[7] Since then, the phenomenon has been looked at from various angles. An inconclusive investigation as part of the Royal Commission did not unequivocally link the black mist to deaths and illness. Indeed, no investigation has been conclusive. The facts of the case are perplexing from beginning to end, and entangled throughout with speculation and surmise. According to Heather Goodall, the bomb test authorities

> had been warned that the bomb should not be fired in the existing conditions in which the wind was blowing in the one direction from the ground up to very high altitudes, because the explosion's plume of radioactive smoke and particles would not disperse. Instead, the meteorologists predicted, it would stay concentrated, drifting north-west in a dense low cloud to pass directly over the known small communities of mainly Aboriginal Australians at Mintabie, [Wallatinna] and Welbourne Hill.[8]

These predictions were accurate and, tragically, they were ignored. Many other imponderables remain. What exactly was the black mist made of and was it capable of killing people?

Ray Acaster gives one of the most persuasive accounts. Acaster was a British meteorologist who came to Australia to work at Maralinga in 1957, and later settled in the country. He helped prepare A32, the British document that set the meteorological firing conditions for Totem and he was one of the meteorologists who warned that firing an atomic weapon in the prevailing conditions at the time of Totem was risky.[9] Several decades later, he applied a critical eye to the black mist, based on his professional experience. He contends that the emergence of the black mist accords precisely with the meteorological conditions at the time of Totem I:

The presence of moisture was confirmed by films taken of the first few minutes of the explosion by high speed cameras. They clearly showed the formation of what is known as a Wilson cloud – when the uplifting of the air by the explosion causes the moisture to condense into visible cloud, albeit for a few seconds only.[10]

Acaster noted that a warmer layer of air was present about 250 metres from the ground, and the most persistent Wilson cloud, a relatively transient puff of condensation, formed just below this layer. The combination of meteorological conditions and terrain was just right for an atomic explosion to create a mist or fog.[11] This possibility had not been factored in by the British authorities, in the early stages of knowledge about atomic weapons and their effects.

Acaster believes that all the physical factors were in place to create this strange phenomenon:

> The Black Mist was a process of mist or fog formation at or near the ground at various distances from the explosion point ... Radioactive particles from the unusually high concentration in the explosion cloud falling into the mist or fog contributed to the condensation process. The process of mist formation, dispersion and re-formation with some sideways divergence would have resulted in a large area covered, not necessarily continuously, by a slow moving 'mist'. The radioactive particles in the mist or fog became moist and deposited as a black, sticky, and radioactive dust, particularly dangerous if taken into the body by ingestion or breathing.[12]

This awful consequence of Totem I could have been foreseen, if all the meteorological and weapon factors had been properly considered. The scientific evidence for the formation of a fog or mist is strong, and these days not seriously doubted. Also, the eyewitness accounts of this elusive, roiling juggernaut are consistent and harrowing. We still don't know, though, precisely what was in the mist and therefore exactly what damage it could do to the human body. Even subsequent medical studies have not conclusively determined what the mist wrought, in terms of population illness and death. One reason is that the health characteristics of Aboriginal populations in the area were almost entirely unknown. In other words, there were no baseline data, and without these, you can't say for sure what the introduction of a hazardous element has done. At the time of the Emu tests, births and deaths of Aboriginal people were not registered, 'nor was there a legal obligation to do so'.[13] As stories about the harm caused by the atomic tests started emerging in mainstream Australia in the early 1980s, a senior official in the Department of Aboriginal Affairs reflected that there 'was very little Government sponsored field activity directed towards Aboriginal affairs or welfare in remote places in 1953'.[14] As historian Heather Goodall said, 'they didn't know who was there let alone how well they were'.[15]

Because so little information is available, establishing the facts remains difficult. In 1980, Dr Trevor Cutter of the Central Australian Aboriginal Congress was head of the Alice Springs Aboriginal Health Service. In May that year the Adelaide *Advertiser* reported his claim that between 30 and 50 Aboriginal people had died from black mist contamination, with a further 1000 seriously affected.[16] Cutter said that he had been told these numbers when he had recently visited the area.

Reports by Cutter and others prompted the South Australian Health Commission to initiate a study of the possible effects of atomic testing radiation on the Aboriginal populations of the area.[17] The authors of the 1981 study acknowledged that it would 'not provide proof of any past or prevailing effects of radiation from atomic tests. A study with this objective would be extremely complex and would need to be undertaken on a large scale'.[18] Not enough was known about what happened when the population was exposed to test radiation and toxins, and later authorities have always had to piece together accounts based on incomplete data. Of course, this was in the interests of those responsible for the tests, whether or not it was consciously engineered that way.

Predictably, the South Australian Health Commission report came out with inconclusive results. 'On balance, the findings provide no clear evidence of a developing trend in disease incidence among Aboriginal populations which could be related to radiation exposure.'[19] The report went on, 'there is a notable absence of information on disease rates for Aboriginals in general, which complicates the selection of suitable standards for comparative study'.[20] Nevertheless, the study did turn up a cluster of five cancer deaths at Yalata in 1979–80, a large number in a small population, although this related to Aboriginal people most likely displaced by Maralinga, rather than those affected by Totem.[21] Beyond that the study declined to go.

Ahead of the Royal Commission, Charles Kerr, professor of preventive and social medicine at the University of Sydney, led an enquiry that considered the available evidence on the black mist, and raised concerns that were later taken up by the Royal Commission. The Kerr report acknowledged that it could not dismiss the possibility that the 'black mist' may have represented a deviant pattern of fallout deposition from the first Totem test. Persisting beliefs among the Pitjantjatjara people of the harmful effects of 'black mist' and related phenomena have not been fully investigated. The Committee concluded that the matter required further examination.[22]

The dearth of evidence and of medical and other records that might have provided clear evidence of the harm caused to Aboriginal people has always been a major obstacle in resolving the issue. The Royal Commission was simply not equipped to conduct a thorough investigation of the health effects or the composition of the mist. Geoff Eames, counsel for the Aboriginal people, said in opening remarks at Marla Bore,

> [we] will not be in a position and nor will we expect and nor will the Aboriginal people expect that your Honour and the commission be in a position to make a positive finding that the deaths that you will hear referred to tomorrow [at Wallatinna] by the Aboriginal witnesses are indeed related specifically to the Totem test. We believe ... that the precise question is one which may well have to be resolved in civil proceedings at a later time when the specific individual issues will have to be resolved.[23]

Both Geoff Eames and his co-counsel Andrew Collett saw it as part of their job to gather information about all potential Aboriginal claims while they were preparing their case for the Royal Commission. When it was over, they identified every Aboriginal person who had given evidence or had been the subject of evidence. They also found anyone identified by an Aboriginal organisation who had asserted that they had been injured by the British tests. Collett and Eames then formulated a civil claim on behalf of these people that was put to the Commonwealth. Eventually, the claim was accepted and compensation paid to all claimants without having to go to court. Nevertheless, the trauma left by the black mist remains and there has been no closure.[24] Although Eames asked for, and received, an open verdict on the black mist at the commission, it caused pain at the time, and still does.

William Penney was asked about the black mist when giving testimony to the Royal Commission in London. He said that he had only recently been made aware of the phenomenon: 'I did not hear at the time, nor did I hear in the next few months. The first that I ever heard of it was perhaps two years ago [1983] when I read it in the British newspapers'.[25] When asked if he had had any opportunity to investigate for himself what it might be, he said: 'No. I thought about it for a few minutes, and I could not match it to anything that I could imagine. Having got that far, I did not do anything'.[26] His energies and attention were directed elsewhere at the time of Totem, and far away from Emu Field in his later years.

Scientists at Aldermaston, headquarters of the AWRE, were curious when stories started appearing in the Australian media from 1980. According to Lorna Arnold and Mark Smith, an investigation that examined meteorological and radiological data from Totem using computer-based analysis

> concluded that the sites at Wallatinna and Welbourn Hill did receive fallout from the test on 15 October 1953. The report said that debris drawn into the convective boundary layer (CBL) – the layer of air close to the earth's surface – could have been sufficiently concentrated to be visible to

observers, and it might have had an appearance similar to that described.[27]

According to Arnold and Smith, the Aldermaston report nevertheless dismissed the idea that any adverse health events could have arisen from the phenomenon.[28] A later report by the Australian Ionising Radiation Advisory Committee (AIRAC), the infamous AIRAC 9 report considered by many to be a whitewash, reached the same conclusion. AIRAC was the successor to the Atomic Weapons Tests Safety Committee. The Royal Commission demolished the credibility of AIRAC 9:

> It would be a conventional objective approach to such a question [of the black mist] to collect, narrate and analyse the available evidence before seeking to express conclusions. AIRAC failed to do this. It was content to look primarily at media accounts. It never sought out anyone with first-hand knowledge.[29]

The AIRAC 9 report was tabled in federal parliament in 1983. It asserted that the black mist could not have caused sickness or death to people at or near Wallatinna, although it conceded that it had apparently caused eye problems, rashes, coughing and diarrhoea.[30] AIRAC 9 asserted: 'All of the evidence for health impacts before AIRAC was anecdotal and there was no corroborating evidence available on the association between the health effects/deaths and radiation exposure'.[31] Clearly no serious attempt had been made by AIRAC to analyse the allegations from Aboriginal witnesses.

Scientists continue to take up the challenge of trying to understand a phenomenon with no pinpointed cause. Dr Geoff

Williams, one of the most knowledgeable of scientists who worked on the two British atomic test sites, investigated the facts of the black mist with colleagues. Williams was a radiation safety specialist from the ARL, and has since retired from its successor organisation, the Australian Radiation Protection and Nuclear Safety Agency (ARPANSA). The study, which published its findings in 2017, took a scientific approach to available knowledge, reviewing what was known about the black mist since it first started receiving scrutiny in 1980. The report concluded that if the black mist was radioactive, the radioactivity was 'well below the level at which acute radiation effects are observed'.[32] The study noted a wide range of possible ways for the cloud of material from Totem I to affect people in its direct path:

- inhalation of radionuclides [radioactive atoms] in the cloud;
- ingestion of radionuclides in the cloud;
- external exposure to gamma radiation from radionuclides in the cloud (immersion);
- external exposure to beta radiation from radionuclides in the cloud (immersion);
- external exposure to gamma radiation from material deposited on the ground;
- external exposure to beta radiation from material deposited on the ground;
- exposure to gamma radiation from material deposited on the skin, and
- exposure to beta radiation from material deposited on the skin.[33]

The first two methods could only have occurred as the cloud rolled over people, during a period of about 10 to 30 minutes. The other forms of exposure potentially could have occurred well after the cloud had passed. Any cloud forming in the atmosphere after Totem I's detonation would include a mixture of fission products such as radioactive strontium, material containing soil and steel from the vaporised tower affected by the fast neutrons emitted by the blast, and unburned plutonium.[34] The Williams paper notes that there 'would have been a wide range of particle sizes in the material (debris) injected into the atmosphere by the explosion, and many of the radionuclides would have been attached to this material'.[35] The neutrons that escape at high velocity from a fission explosion, phenomena measured by Ernest Titterton at Emu, collide with other matter such as soil or metal. Was this the source of the toxicity of the black mist? Williams and his colleagues did not discount this.

Most scientists, while not ruling out chemical toxicity, are sceptical that the mist was sufficiently radioactive to cause illness. Williams raised another intriguing possibility: that the radioactive isotope cobalt-60 may have caused the radioactivity in the black mist. At Maralinga pellets of cobalt-60 were deliberately (and underhandedly) added to a bomb test as a (failed) experiment in detecting the explosive energy of the blast. At Emu the steel towers contained the stable isotope cobalt-59. When the Totem device released a large quantity of neutrons upon detonation, some of these neutrons will have interacted with cobalt-59 in the towers to produce the much more dangerous radioactive isotope, cobalt-60.[36] This may have formed part of the black mist: 'there is a possibility that radiation emitted by the radioactive decay of [cobalt-60] produced by neutron activation could contribute to the dose received by the people at Wallatinna'.[37] This hypothesis remains speculative. Certainly radioactive cobalt is dangerous. If ingested, it is absorbed into the tissues, especially the liver, kidneys and bones, and may cause cancer.[38] Sending it out, uncontrolled, into the atmosphere was extraordinarily reckless. Interestingly, the towers at Maralinga were constructed from different material.

The Williams study did not put forward a detailed or conclusive explanation of the black mist phenomenon. In addition to its speculation about cobalt-60, the paper also suggested the possibility that the phenomenon wasn't directly from the bomb at all, but rather 'a smoke plume produced by a fire that started as a result of scrub being ignited by hot debris from the fireball'.[39] The paper's conclusion joins Aldermaston and AIRAC in doubting radioactivity in the mist caused sickness, while acknowledging 'there is no reason to doubt the evidence that these same Aborigines experienced acute and chronic health effects around the same time'.[40]

Similar phenomena have been reported elsewhere when atomic weapons were exploded, including at Maralinga. The Royal Commission noted that Walter MacDougall had sent a telegram to the Maralinga director on 2 October 1956 in the wake of the first Maralinga test to report being told by workmen that a black cloud had detached itself from the main atomic cloud before rejoining it later: 'At night when they were in bed particles of sandy dust were hitting their canvas camp sheets very similar to raindrops'.[41] The Royal Commission also noted analogous accounts from tests in the United States.[42] There are reports of a black rain that fell soon after the bomb dropped on Hiroshima in 1945 too.[43] Together these suggest that the black mist of Totem I was not unique.

However, the science on the black mist is not in, and may never be. The ongoing secrecy around the exact nature of the

Totem tests contributes to the lack of understanding. All of that is immaterial in the face of the suffering and fear induced by the mist. One of the most eloquent witnesses of the black mist was the distinguished Yankunytjatjara elder Yami Lester, who lived through its terror as a 12-year-old child. As a young lad he enjoyed the bounty of a rich natural environment and the freedom of living in his ancestral lands, but after Totem I, nothing was ever the same. That October day, he felt the ground shake ominously as a prelude to the arrival of the darkening cloud.[44] Almost immediately, one of his uncles got blisters over his body.[45] There was no medical clinic at Wallatinna, and by the time his uncle saw a doctor it was too late and he died. Lester and his family moved away from the area and he did not return for many years. As a teenager, Lester was a stockman at Granite Downs for several years, until his failing sight made it impossible. He lost sight first in his left eye in 1957 when he was 16, and then in his right in the mid-1960s. Although shaken by the dislocation and from the trauma caused by the emanation from Emu, he always considered Wallatinna to be his place, the place where he belonged and he would spend some of his later years there, as would his children.

> So I didn't think about it for a long time. Never mentioned it to anyone. I don't know what sort of illness it was, and I still don't know. But it stayed in my mind somewhere. I only thought about it years later, that maybe what happened to my uncle had something to do with what happened to my eyesight.[46]

Lester sought medical advice in Adelaide, where Dr David Tonkin, an eye surgeon who later became the premier of South Australia, operated on his right eye in 1965. Tonkin thought it worth trying a corneal graft, but the treatment was ultimately unsuccessful, given the extensive damage to the eye.[47] Tonkin gave evidence to the Royal Commission that the blindness had possible causes other than exposure to the black mist, and he could not himself see how ionising radiation could have been the cause, based on what may well have been a narrow reading of Lester's circumstances.

After Yami Lester went blind, his life changed forever, but he adjusted. He became a prominent community leader, heading up the Institute for Aboriginal Development in Alice Springs. He ran a radio service and was often on the radio himself. As he recounted in his autobiography, one day in 1980 he was listening to the ABC Radio current affairs program, *AM*, and heard an interview with Sir Ernest Titterton, comfortably ennobled and enjoying his life as a senior academic at the ANU. Titterton was an occasional acerbic media commentator whenever anyone wanted to talk about Britain's atomic test program in Australia. Lester was at home, sick with the flu.[48] He was only vaguely listening to the still-British-sounding Titterton. Then Titterton was asked, 'And what about the Aboriginal people' and replied, 'Oh, the black people were well looked after. We had two patrol officers in the area.'[49]

Lester stopped and gave the interview his full attention. He recounts in his autobiography, 'I said to the radio: "Bullshit!"'[50] He brooded on the interview all day, and memories kept rushing back to him of his childhood experiences and trauma. 'I was a kid then and didn't really understand. Well, I still didn't understand, but what I remember wasn't anything like Sir Ernest Titterton's story. Someone had to answer that person ... So what was I going to do?'[51] He contacted Rob Ball, a journalist from the Adelaide *Advertiser*. Ball subsequently wrote a story published on 3 May

1980 that was the first mainstream public disclosure of the black mist phenomenon. Lester soon became involved in answering the misinformation about the atomic tests in Australia, a role that continued until his death in 2017. He also contacted Dr Trevor Cutter to talk about the painful memories that were returning.[52] As he said in his book, 'I am not a medical man and I don't have any records, so I can't prove anything, but it's true: all those [Wallatinna] people have eye problems'.[53]

In the latter years of his life, Yami Lester was a staunch, compelling and persistent voice for the Aboriginal people harmed by the Totem tests and the later tests at Maralinga. His name is now intrinsically linked with the black mist and the atomic tests, and with his pioneering advocacy for the truth of what happened. Two London-based journalists, David Leigh and Paul Lashmar, interviewed him for the *Observer* newspaper in 1983. In their story, republished by the *Bulletin* in Australia under the title 'Forgotten Victims of the "Rolling Black Mist"', Lester said:

> I looked up south and saw this black smoke rolling through the mulga. It just came at us through the trees like a big, black mist. The old people started shouting 'It's a mamu (an evil spirit)' ... they dug holes in the sand dune and said 'Get in here, you kids'. We got in and it rolled over and around us and went away.
>
> Everyone was vomiting and had diarrhoea, and people were laid out everywhere. Next day, people had very sore eyes, red with tears, and I could not open my eyes ... Five days after the black cloud came, the old people started dying.[54]

Lester's testimony was always consistent and vivid. His statement to the Royal Commission, and his in-person testimony at Marla Bore in 1984, never faltered. The causal link between the black mist and his blindness is not entirely clear, as his autobiography concludes, 'And I never did find out if one of those bombs made me blind'.[55] The connection between Lester's blindness and the mist has been amplified over the years, and many now accept that he was blinded by the black mist. Certainly, his blindness followed soon after a traumatising experience. Perhaps the black mist permanently damaged his health, leading to a cascade of other health problems that robbed him of his sight. As Lester said in his autobiography, 'I am still angry for what they did'.[56] His mother, Pingkai (also known as Pingkayi), and step-father, Kanytji, also gave evidence to the Royal Commission, at Wallatinna. Through an interpreter, they testified to the sickness and death that befell their community after the strange black frost-like material settled on the ground and vegetation.[57]

Another compelling witness at the Royal Commission was Lallie Lennon, an Anangu woman who followed Lester in giving evidence at Marla Bore. Lallie Lennon had gone through the ravaging measles epidemic of 1948 when she lived at Oodnadatta as a young mother. She and her husband Stan later moved to Mintabie, where Stan prospected for opals. After breakfast on 15 October they heard and felt 'a very long rumble'. Then smoke followed, not like a dust storm but something different; it smelled of gunpowder and stuck to the trees.[58] Lallie Lennon and others at Mintabie used to eat the honey from some of these trees and they all agreed not to eat it after the strange material stuck to the branches. Even so, Lallie, her children and the others with her all got sick with diarrhoea and flu-like symptoms, and some developed rashes and sore eyes.

Soon after, the family went to Coober Pedy to recover.[59] Lallie Lennon's ongoing skin problems were so severe, she looked like she had rolled in a fire.[60]

The stories of the survivors of the black mist are relentlessly harrowing. One eyewitness, Ellen Giles, a white woman, presented a statement to the Royal Commission in which she painted a grim picture of an event that was frightening and inexplicable. Ellen Giles was proprietor of a station at Welbourn Hill, where she lived with her husband and three children. She hired young Aboriginal girls as cleaners. One came in

> and she said Mrs, it's gone, all gone everywhere. It wasn't like a wind, it was just sort of coming in. We went outside and we touched the doors, everything, and it was horrible where the thing had gone over the house. That killed our trees, all that had been there years ago when the Giles came back from the First World War.[61]

Another survivor who testified was Auntie Merdie Lander, whom Yami Lester described as 'a wonderful witness'.[62] Lander, who was not Aboriginal, was interviewed at Marla Bore, just before Lester. At the time of the black mist, she and her husband were living at Never Never just to the west of Welbourn Hill station. Her descriptions are particularly vivid:

> Although it was the colour of a rain cloud, darkish, it did not have the compact, rolling look that a rain cloud would have. It was just a sort of a mass, and it had at the top of it, which was most unusual, like a banner, I would describe – as a banner that stretched ... upward.[63]

She and her husband were living in a caravan, and she was used to the dust storms that would come through often and which she could clean up with a whisk of a duster. Not so the black mist. '[T]his could not be flicked off with a feather duster. It had to be wiped off with a damp cloth. It was quite sticky.'[64] Lander went on to suffer skin problems after the black mist descended on her home, including scaly patches and lumps on her arms, legs and face.[65]

In October 1983, the community at Wallatinna held a commemoration, a kind of re-enactment, of the black mist 30 years before. So much had changed. The rise of land rights activism from the 1960s meant that Aboriginal communities were starting to find their voice, and often expressed anger at the abuse they had received. A video recording of the commemoration was taken to London by Yami Lester and his wife Lucy, accompanied by Pitjantjatjara Council lawyer Richard Bradshaw in June 1984. Bradshaw, an Englishman, moved to central Australia in the early 1980s and was captured by the land, its people and its issues, and has taken on land rights cases ever since.

The video must have been compelling, although no copies seem to have survived. As Bradshaw recalled, 'People weren't rude about it. Their perceptions were – they weren't going to dismiss it, at least to our faces'.[66] The trio attended many meetings, including one with Geoffrey Pattie, the minister for defence procurement in Margaret Thatcher's government, along with a range of civil servants and others with an interest. Bradshaw recounts how Lester brought his large tape-recording device, a 'boom-box', purchased during the stopover in Singapore on the way to London. He placed the bulky recorder on a narrow, rectangular table during meetings with Ministry of Defence and AWRE representatives, making the not-unreasonable claim that his blindness made it necessary to record the conversations. The meeting participants were wary, but out of politeness acceded to his request. Bradshaw recalls one particular recording Lester made of Ministry of Defence officials: 'That was a remarkable thing because there were certain admissions that got made in the course of that meeting at least. These were very useful, not because they were truthful but because they weren't truthful.'[67] Lester was nothing if not resourceful. The extensive publicity that accompanied their trip to Britain and the lies they were fobbed off with in Whitehall gave further impetus to the case for a Royal Commission.

Dr Heather Goodall, a researcher attached to the legal team who represented the Aboriginal people at the Royal Commission in the mid-1980s, was struck by some interesting anomalies in witness interviews in the lead-up to the enquiry. In an article published in 1992 she examined in a nuanced way the complex interweaving of the black mist with other traumatic events in the life of Anangu people. In doing so, she suggested a 'divergence between history as produced for the law and an historical appreciation of the place which a catastrophic event may assume in community memories'.[68] Goodall was particularly interested in the differences in the accounts of the black mist gathered by the Royal Commission from people who were in the Wallatinna area compared with those of people at or near the Ernabella Mission about 250 kilometres to the north-west. Goodall and the legal team realised that the Ernabella witnesses were remembering aspects of the 1948 measles epidemic, not the black mist of 1953. The two events had become entangled. Because of these discrepancies only evidence from the Wallatinna accounts went forward to the hearings.

For instance, Kunmanara (his Indigenous protocol name;

his actual name was not permitted to be used), patiently told Goodall about his experience at Ernabella when she interviewed him before the Royal Commission. Kunmanara was convinced that the Emu tests and the deaths caused by the measles epidemic were connected. In his youth, he had been camping slightly away from the mission with other young men, preparing for their initiation ceremonies. An Ernabella missionary brought them food and told them of the measles outbreak at the mission. The boys were forced to watch in horror from a distance as their loved ones sickened and, in many cases, died, their bodies brought out on stretchers. 'The boys sitting on the edge of the hills could not see who was dead, but they saw the bodies carried away in wheelbarrows, sometimes two and three at a time, for hasty burial. Chilled with fear, the *nyiinka* [young men] could do nothing except watch.'[69]

Over the years, as Kunmanara drew upon his cultural knowledge and his eyewitness experience to make sense of this horrific time, he looked for a plausible explanation. His terrible experiences in 1948 started to accord with the idea of poisoned land and this was reinforced by the way Walter MacDougall described the bomb tests. MacDougall had little time to warn Anangu before Operation Totem. He visited Wallatinna just before the Emu tests, but he didn't get to Ernabella until the following year, 1954, when he warned of the upcoming Maralinga tests. He attempted to provide a comprehensible frame of reference to something completely unfamiliar to the people of the area. According to Goodall, 'To explain the tests, he drew sketches which were not of the mushroom clouds which have come to symbolise nuclear explosions, but instead were of metal bomb cases, which he accompanied with descriptions of various second world war campaigns'.[70] Goodall later explained that MacDougall appeared to be desperately worried that Aboriginal people would head south into the atomic test areas and 'he was pulling out everything he could think of to try and dissuade them'.[71] MacDougall's instincts were good. 'He didn't appear to have any sense of institutional loyalty to the Weapons Research Establishment. He seemed to be very focused ... on the potential danger to Aboriginal people.'[72] Goodall wrote that it 'seems that MacDougall himself, between 1953 and 1956, did much to link the physical properties of the bomb to warfare, to the measles epidemic, and to Pitjantjatjara concepts of sorcery, *mamu* and death'.[73] At the time, MacDougall advised his superiors at Woomera that the best course of action was to use 'their own beliefs and fears of evil spirits and invisible avengers, convince them that that area [Emu Field] is no safe place for them'.[74]

As Kunmanara processed his own traumatic experiences in relation to the particular way MacDougall described the atomic weapons tests, he concluded that the black mist must have come to Ernabella. As Goodall eloquently put it:

> While his history was not directly relevant to the physical effects of the bomb tests, it nevertheless speaks to us about how the nuclear tests came to assume metaphoric significance for Anangu over the years, perhaps allowing them to express their anger at all the other devastation brought to their lives by invaders with whom they could rarely openly express hostility. Kunmanara's testimony suggests his struggle to make sense out of the whole society and culture which had invaded his lands and brought death by fatal diseases as well as having created and exploded their bombs.[75]

While there is no suggestion that the black mist itself was a myth – the black mist was all too real – some of the stories about where it went have a more complicated relationship to the Emu Field story. Goodall compares this with work done on the effect of atomic testing in India, at various times since 1974, underground and adjacent to an army base.

> Only recently have people in the villages around been able to talk about their concerns, the cancers and the numbers of deaths. In many ways it is like the situation in central Australia. There was no baseline study, not enough adequate statistics, but a lot of health damage that people suffered. The possibility of identifying actual damage is very difficult, indeed unlikely. What this is telling us is for all this time, since 1974, these people [in India] have been worried sick. Every time someone has died there has been a question mark – was this because of the detonation? Nobody knows and nobody ever will know, and people have been worried sick.[76]

These same fears have pervaded the Aboriginal communities around both Wallatinna and Ernabella.

Similar narratives of anxiety can be found among other communities in the vicinity of Emu Field, including Coober Pedy. The mist did not reach Coober Pedy physically, but the accounts of poisoned land, sickness and death certainly did. As Goodall points out, the weight of guilt and anxiety that people carry, even forms of post-traumatic stress disorder, need to be recognised. The black mist was not just a physical phenomenon but a psychological one that shows the damage in populations haunted by the possibility of unknown and unquantifiable harm. In Goodall's words, 'this has obviously been intolerable for everybody'.[77] The failure to inform Aboriginal people about what was being done in their homelands may be the biggest betrayal of all in the story of the British nuclear tests in Australia.

At the commencement of the London hearings of the Royal Commission, counsel for the Aboriginal people, Geoff Eames, made a poignant plea:

> our clients' attitude has been from the outset and continues to be that their principal concern is and has always been to remove anxieties that they have had in the past, to remove anxieties that might occur in the future ... If the anxieties that Aboriginal people feel about this test program are proved to be groundless, they will be delighted to recognise the fact.[78]

Like so many who lived in the vicinity of Operation Totem, the burden of uncertainty and anxiety weighed heavily, and still does.

Chapter 6
SECRETS AND SAFETY LIES

Once the contaminated particles have been thrown into the air they will be dispersed by the wind and fall to the ground under gravity. Where they fall they will contaminate the ground and if this contamination is heavy enough it could conceivably present a danger to the health of any person in the near vicinity.

High Explosives Research report A32.

[A] visitor is liable for prosecution if he discussed any aspect of his visit with any unauthorised person (which includes close relatives).

Official guide for visitors to X200.

Yes. From a security angle the explosion of that bomb - was a perfect job. Even the tiny Russian dwarfs said to have been in the vicinity, peeping out of kangaroos' pouches, couldn't have learned much.

Ronald Monson, *Daily Telegraph*.

Secrets and lies pervaded Operation Totem, perhaps the most furtive British atomic tests held in Australia. The Australian Government was kept at bay and, characteristically, did not ask hard or uncomfortable questions. The secrecy of Totem was not just for strategic reasons. It was used to cover up how little thought Britain gave to the consequences of its testing, the harm it visited upon Aboriginal populations and military personnel and even Australian civilians as far away as Townsville, who were entirely unaware they were in the path of radioactivity from the Totem shots. At the very least, security-cleared senior members of the Australian Government should have been informed of the full reality of the tests, but they were not. They should have asked more questions too. The ripple effects are still evident, and the continuing impenetrable secrecy looks even less reasonable today.

The theme of secrecy and lies was established from the beginning. Official restrictions denied the fact of the tests for as long as possible, using the Woomera rocket range as convenient cover. Lies were needed to sustain the secrecy. Ben Cockram from the UK High Commission in Canberra wrote to Prime Minister Menzies in March 1953 spelling out the British desire to hold information about the forthcoming tests close: 'we wish to defer as long as possible disclosure that the tests are to take place ... We hope to be able to dispose of speculation until quite close to date of test by reference to rocket range activities'.[1] The message was crystal clear: Australia was expected to keep information about the tests from its own citizens until it was nearly time for them to begin. The Australian Government was happy to oblige.

Meanwhile, plans were developed rapidly in that madly busy year of 1953. As part of the build up to Totem, William Penney called for the preparation of safety requirements for firing the two Totem bombs. The guidelines were duly prepared, only to be largely ignored. They were contained in a top-secret document known as A32, prepared in May 1953 by four meteorologists working for the AWRE, with the official title 'High Explosives Research Report No A32 – Materials and Physical Research Division (HER) – Airborne Contamination at the Totem Trial'.[2] It was written just as the ambiguously named HER was adopting the much franker name of Atomic Weapons Research Establishment (AWRE). A32 is perhaps the starkest indication of all that the British were flying blind, without sufficient experience or knowledge to be able to predict the outcome of their bomb tests. The predictions contained in A32 were based on a range of inaccurate assumptions intended to manage risk but which instead greatly added to it. These assumptions made its statement that 'we have shown that there is in general no risk to any of the civilian population of the region' wrong.[3] The document's claim that 'we have tried to over-estimate the danger rather than the reverse' turned out to be false.[4] Prior to Totem the Blue Danube device had only been tested once, in Operation Hurricane a year earlier. The Totem devices were smaller and used different nuclear fuel, so they would inevitably perform differently in numerous respects. Like all scientists, Penney was of a curious disposition. He wanted to find out what would happen to his theoretical calculations in the real world. Unfortunately, this placed lives at risk. Penney colluded in keeping those risks secret both at the time and into the future.

Some of the A32 assumptions concerned the rise of the atomic cloud upon detonation. 'As the mechanism of the rise of the mushroom and the consequent distribution of the contamination in the air is not yet understood in detail we must make some assumption [sic] about the initial distribution in the mushroom head and stem.'[5] The authors based their safety predictions on what they had been told about the size of the device. Since both devices, upon detonation, far exceeded yield predictions, the associated predictions of cloud height and behaviour were incorrect.

Andrew Collett, a member of the legal team representing the Aboriginal people at the Royal Commission, discovered A32 among over 500 British documents made available to them several days before evidence was taken in London in January 1985. The document contained a complex and technical series of revelations that had to be dealt with on the run. Its significance, however, was immediately clear to Collett, who shared it with his co-counsel Geoff Eames. Their final submission argued A32 showed that, '[e]ven using the criteria for safety which the scientists adopted ... Totem I should not have been fired. The firing clearly breached the injunctions set down by A32 when an adjustment from a 5KT to a 10KT yield weapon was considered':

> No adequate explanation has been given to the Commission for the fact that Totem I was fired contrary to the provisions of A32. The seriousness of the disregard for the terms of A32 is compounded by the fact that ... A32 itself had underestimated the extent of contamination of populated areas which would result from firing weapons in the conditions specifically proscribed by its authors.[6]

Despite its flaws, making use of A32 would have been better than disregarding it completely. The submission further argued that the meteorological conditions at the time of firing were not suitable: 'Lord Penney when first called to give evidence ... refused to concede that the wind conditions described ... almost precisely mirrored the conditions specifically warned against in A32'.[7] Later, Penney came back to the stand and grudgingly conceded that Totem I should not have been fired. However, he claimed that the risk to Aboriginal people was '20 to 1 against a single person' contracting cancer from the fallout.[8] What a way to test a hypothesis.

Penney admitted at the Royal Commission that he based his Totem safety calculations on incomplete information about American tests in Nevada. Because the Americans were constrained in what they could share with the British, some of Penney's calculations were based on surmise rather than clear scientific information.[9] He effectively took a map of Nevada, worked out the distance between the bomb sites and places of habitation, and extrapolated.

Ernest Titterton was outraged when Penney conceded under cross-examination that Totem I should not have been fired in the conditions evident on 15 October 1953. The media reported him in 1985 saying it 'sounded like the answer of a very tired man', someone who 'would agree to anything at the end of a very lengthy session of questioning ... I would have to say that he is wrong'.[10] Titterton never eased up in his conviction that the British tests in Australia were entirely beneficial and laudatory, and no harm was done.

Both Operation Hurricane and Operation Totem were carried out before the creation of the AWTSC. This committee was instigated by the Australian Government in 1955 to act as a safety mechanism reporting to the Australian minister for Supply. It was highly problematic and did not provide reliable information to the Australian Government, particularly after 1957, when it was run by Titterton. Nevertheless, it did have safety oversight responsibilities that were not in place for the early British tests. The AIRAC, which superseded the AWTSC in 1973, pointed out to federal science minister David Thomson in March 1980 that the Totem tests were not monitored by an Australian organisation 'and Australia did not share in their control'.[11] Similarly, there was no system of radiation monitoring stations around Australia until AWTSC established them.[12] The lack of any Australian mechanism for control of such risky tests is indeed striking. At the time of Totem, Australia was only allowed 'observers', an ambiguous term at best. Ernest Titterton, one of the observers, along with Leslie Martin and Alan Butement, conceded in his submission to the Royal Commission that observers had no formal terms of reference. He said he was told during a discussion with Menzies that:

(a) The Observers would have no formal responsibilities for the conduct of the operation;
(b) The Observers should lend every possible support and aid to Dr Penney's team carrying out the tests;
(c) The Observers should look after Australian interests in all possible ways. Especially they should satisfy themselves that any bomb fired should be under conditions such that there would be no danger to Australians, or to fauna and flora, from the expected fallout;
(d) The decision to fire would be a British responsibility alone, but we could expect our views to be given careful consideration.[13]

The prime minister specifically mentioned Aboriginal people in relation to point (c), he said, but there were no other rules.[14] Because Titterton undertook nuclear physics experiments at Totem, he had access to highly classified information about the weapons, so he knew much more about the tests than either Martin or Butement, both of whom were kept at arm's length.

When the Australian Totem Panel met in Melbourne on 24 June 1953, the group, led by Jack Stevens, considered an independent safety evaluation carried out by Titterton and Martin. As the meeting minutes laconically relate, 'Professors Titterton and Martin had informed the prime minister that no habitations or living beings will suffer injury to health from the effects of the atomic explosions proposed in the Totem trials'.[15] Within hours of the detonation of Totem I, that assurance proved to be hollow.

Martin sent a letter to Menzies on 17 June 1953, also signed by Titterton, that gave the Australian prime minister a quick lesson in the effects of atomic weapons:

> The atomic bomb resembles a conventional bomb in that its destructive effect results from the liberation of a very large quantity of energy in a restricted volume and in a very short interval of time. The energy released by a typical atomic bomb is however very much greater, perhaps thousands of times greater, than for a conventional bomb. This results not only in a very violent shock wave but also in the emission of thermal radiations of exceptional intensity. Coupled with these radiations are penetrating gamma rays and neutrons which are dangerous to living organisms. The atomic bomb differs from a conventional bomb in another aspect which is particularly important in planning atomic explosion trials. The volatilised products of the explosion become airborne, and being radioactive may on falling out create large areas of contamination which are biologically dangerous.[16]

Exactly what Menzies made of this sobering information about weapons being tested on Australian territory is not directly known. However, Menzies did talk publicly and often about how safe everything was and how no-one should be concerned. Martin and Titterton were key to the government's relaxed attitude. Martin assured the prime minister that the predictions about Totem explosions made by Penney and his colleagues were rock-solid and exhaustive, based upon what they learned from Hurricane, even though the lessons from Hurricane were incomplete. 'The performance of the weapons to be tested in the Totem trials can be predicted therefore with confidence', said Martin.[17] This belies the fact that the British were still in the early stages of atomic testing and did not know the implications of detonating an atomic device in open territory. Martin and Titterton gave the go-ahead for an uncontrolled experiment on people, animals and terrain.

Things got complicated briefly when Martin and Titterton were given access to A32, and they became alarmed. As shown in the Royal Commission Report, Allan McKnight of the Prime Minister's Department reported on 12 June 1953,

> our two Professors had told us orally that safety of Test was amply provided for. However, after conferring in Melbourne yesterday, they withdrew their opinion and have cabled Penney raising a query which I am told is basic to the whole safety appreciation. You will appreciate that until Australian Government receives firm advice from Martin and Titterton as to safety, there can be no question of an announcement and my hopes for a release next week can therefore be discarded.[18]

In their note to Penney, reproduced in the Royal Commission Report, Martin and Titterton raised the issue of upper-level winds and predictions of the maximum height of the cloud. They were concerned that unexpected local rain could see fallout drop on an oblivious Melbourne.[19] Stevens sent a secret cable to Elmhirst quoting Martin, and requested the message be passed immediately to Penney. It contained the chilling line 'Worried about Melbourne in event of unexpected local rain'.[20]

Penney wrote back assuring the two professors that there was no risk, and they were soothed. With the agreement of Titterton, Martin was happy to reinstate their original advice that all would be well:

> On the basis of the information made available by Sir William Penney we are able to assure you that the isolation of the site of the Totem trials precludes any possible damage to habitation or living beings by the 'shock' wave, thermal radiation gamma rays, and neutrons. A study of the meteorological conditions at the site over past years has established that winds of suitable direction and acceptable strength occur over the period of the trials. Short range and long range meteorological information will be supplied continuously from a number of stations specially established for these trials. It is possible for us to assure you that the time of firing will be chosen so that any risk to health due to radioactive contamination in our cities, or in fact of any human beings, is impossible.[21]

Titterton and Martin made this 'heroic' pronouncement without taking into account just how patchy British knowledge actually was. One of the conclusions of the Royal Commission stated:

> Bearing in mind that the yield given in the planning document was about half that of the actual Totem explosions, the categorical and all-embracing nature of the assurance of safety given by Martin and Titterton gave

legitimacy to the Australian Government's decision to allow the tests to take place.[22]

The Menzies government readily accepted the advice of Titterton and Martin, both of whom were experts who had access to some inside information. The government had no other access to technical information other than what was channelled through these two observers.

British atomic weapons scientists scrambled to increase their knowledge base as their test program expanded. Hurricane was carried out by the seat of the pants by a research establishment still struggling to catch up to the Americans. The only way to get up to date quickly was to tap American knowledge. A team from High Explosives Research visited the United States in mid-1953 to better understand how to monitor the effects of an atomic weapon. Here they learned how to use a mass spectrometer, an apparatus designed to analyse small samples and determine the isotopes.[23] They were also introduced to the technique of measuring the electromagnetic pulse (EMP) of an atomic explosion, a method that would enable more detailed information about the explosion.

The McMahon Act was a stumbling block to this exchange, which meant, as scholar Michael Goodman wrote, '[a]lthough the Americans had conducted monitoring experiments on EMPs, under the McMahon Act they could not provide their results to the British; instead, as John Cockcroft discussed, they "strongly recommended it to our attention"'.[24] The British used backchannel interaction to get around American law. During the visit, the United States was invited to send B29 Superfortress aircraft to Totem, to take cloud samples for analysis and work co-operatively on airborne sampling.

The British unilaterally set measures of risk in connection

with the atomic tests that were outlined in A32. However, these risk standards were not those of the International Commission on Radiological Protection, the body that had determined safety standards for ionising radiation since 1928. Instead, the British went their own way. AWRE health physicists at Aldermaston formulated two standards: 'zero risk' and 'slight risk'. Zero risk was taken to mean 'that quantity of fission products which will cause no measurable effect on the body'.[25] Slight risk was defined as 'that quantity of fission products which may cause some slight temporary sickness to a small number of people who have a low threshold sensitivity to radiation'.[26] The British followed these basic standards throughout their atomic tests in Australia, with some amendments over the years. While the Royal Commission found these standards 'reasonable', it was not convinced by their implementation:

(a) The Totem 1 test was fired under wind conditions that the study in Report A32 had shown would produce unacceptable levels of fallout. Measured fallout from Totem 1 on inhabited regions did exceed the limits proposed in Report A32.
(b) The firing criteria used for the Totem 1 test ignored some of the recommendations of Report A32 and did not take into account the existence of people at Wallatinna and Welbourn Hill down-wind of the test site.[27]

A32, which historians Lorna Arnold and Mark Smith described as intended to be (but was not in the end) 'safely pessimistic', was predicated on the estimate of an explosive yield for Totem I of 5 kilotons.[28] In the event, the yield was 9.1 kilotons. On these grounds alone, Totem I should not have been fired.

While the British authorities were parsimonious with the

information they provided to the Australian Government, the Australian public remained even more underinformed. Efforts were made to either keep information from Australians or distort the information that was provided. Oddly, perhaps, the British authorities wanted to hide the true nature of the Totem tests while simultaneously boasting of their efficacy. They often seemed concerned that Totem's lower yield compared to Hurricane's would be seen as a black mark. They worked to defuse this:

> It is intended to ask Australian Army officers to help with the Radiological Safety Survey teams, which will give them valuable military experience and will reduce the number of people required from the U.K. These officers will obtain information about the degree of contamination involved, but only from low-yield weapons. To prevent them from underrating the U.K. military Atomic Weapons, it is recommended that they should be told that the object of the trial is to determine how the yield of the United Kingdom weapon varies with the weight of the plutonium content. Both weapons will contain less plutonium than the Monte Bello weapon, and one of them substantially less.[29]

There were other risks associated with Operation Totem aside from radiological ones. Anyone based at Emu Field had to confront a range of pressing matters that could affect their wellbeing, and the conditions presented non-trivial health and safety issues for the troops stationed there – especially British troops with no experience of desert life. The booklet they were given upon arrival, 'Welcome to the Claypan', gave some hints. Len Beadell's cartoon on the front cover showed British servicemen in summer issue gear sweating profusely next to an aeroplane on the airstrip while being watched by an amused kangaroo.[30] X200 denizens were warned not to stray from the roads: 'Short cuts through the mulga scrub (known as "bush bashing") can be an interesting experience but it is attended with real dangers and in any event becoming lost is a simple matter'.[31] The whole booklet was decorated with Beadell's drawings (as indeed was the Emu mess hut, which featured two large and, it must be said, insulting caricatures, one of a toffy Brit and another of an Aboriginal man). While the booklet was lighthearted and no doubt intended to be welcoming, there was no denying the seriousness of a place where deviating even slightly from a rough track could be fatal.

It was particularly important that no personnel were near the bomb as it detonated. The safety mechanism to deal with this was the use of 'personnel keys'. Seven actual keys were issued to officers who were responsible for the safety of their men. When the time came for a detonation, each officer had to confirm that everyone in their charge was 'under safe cover' and return their key to the control area, where it was inserted into a corresponding switch-lock on the control desk. The official protocols for the tests spelled out that 'only when all the keys have been inserted and turned is it possible for the control sequence to start'.[32] A master safety link, in the hands of the group leader of weapon operations, provided a way to break the connection between the control desk and the weapon. Once activated, it would return the weapon to its safe condition and stop detonation.[33] There were other methods devised to enable key members of the team to delay or disarm the system if need be, although they do not seem to have been used.

Beyond the claypan, secrecy about the tests was near absolute. While Australian journalists were largely biddable, understandably they wanted to know about British activities in central Australia, as did the general public. However, they only got bland statements, provided at the last minute, with virtually no information. Stonewalling the Australian Government was a more delicate matter. A top-secret Ministry of Supply briefing document written by FC How stated:

> The Committee understands from Sir William Penney that the Australian Government will expect to have rather more information about 'TOTEM', which is to take place in the heart of the Australian continent, than about 'HURRICANE', and he suggests that ... no design details of atomic weapons should be disclosed to any Australians.[34]

He did allow for some information to be shared:

> We are already committed to disclosing to two Australian scientists (Professors Martin and Titterton) details of the contamination so that they can make their own independent evaluation of the hazards. We must therefore offer them information about the approximate yield of the weapon, and an opportunity to witness the bursts. The necessary information will have to be given to them soon, since the Australian Government will need to be armed with their evaluation before there is any possibility of an announcement, or a 'leak', about the tests.[35]

The briefing document recommended that Penney brief Martin and Titterton orally 'about the general principles of the implosive compression and how it is achieved', when he arrived in Australia just before the tests.[36] Inevitably some less senior people would have to be trusted with secret information:

> Australian help will be needed to hoist the weapons on to the firing towers. This task will fall to the Australian Army, and the weight and physical size of the weapon will become known to those who undertake it. It is presumed that all concerned will be Security cleared by the Australian authorities.[37]

All news intended for the general public was to be strictly controlled, however, with the British Ministry of Supply advising that the Australian Government

> should take the lead in judging when to propose that an announcement ought to be made, since speculation is likely to build up more heavily in Australia than in the U.K., and it is the Australian Government which will have to face any political difficulties which may arise, i.e. with the Aborigines Protection Societies.[38]

If these groups had known about the risks, they may have been able to better prepare Aboriginal people for what was to come, but that was not to be. Only the possible 'political difficulties' were mentioned.

Britain sought the involvement of New Zealand in the Totem tests. For Hurricane, the Royal New Zealand Air Force had dispatched aircraft from Whenuapai, near Auckland, to collect samples about 5600 kilometres from ground zero and '[t]hree of the four aircraft collected significant radioactivity'.[39] A top-secret telegram sent during 1953 from the CRO to the UK High Commission in New Zealand, and copied to the UK higher commissioner in Australia, made it clear that fallout from the tests was expected to encroach upon New Zealand:

> We should again like to have assistance of Royal New Zealand Air Force aircraft. Their task would be collection of samples from the upper air drift in an area over New Zealand. If you see no objection we should be grateful if Chief of Air Staff could be approached and asked whether he is agreeable in principle to provision of aircraft for tasks similar to those undertaken after Monte Bello explosion ... You will appreciate that this matter should be treated as highly secret. We wish to defer as long as possible any disclosure that a test is to take place, and even when announcement of this is eventually made, it will be necessary to preserve secrecy regarding details including in particular scheduled date of explosion. Telegraphic correspondence on this subject should be prefixed 'Crusader'.[40]

In the event, New Zealand declined to involve its aircraft in Totem, despite their participation in Hurricane and, later, Operation Buffalo at Maralinga.

Some aspects of the Emu story will probably always remain shrouded in mystery. Secrecy creates an information vacuum. Since Emu Field was especially secret, it has generated many compelling and intriguing stories. Some are unverifiable; some reflect the way Aboriginal communities understand the catastrophic events; some are exaggerations or even deliberate falsehoods. The Aboriginal people of the areas adjacent to the Emu Field test range, and the military veterans bound by duty, have carried anxiety for decades. Their anxiety is not to be dismissed. It is real and it continues.

Rumours and stories concerning the death of Aboriginal people at Emu in the actual bomb blasts themselves remain unproven and are most likely unprovable. Some stories have been discounted, including those of Erron (Terry) Toon, a member of the Maralinga and Monte Bello Atomic Ex-Servicemen's Association. Toon wrote to Bill Hayden, then federal member for Oxley and former leader of the Labor Party, in May 1984, outlining allegations that both the Emu and Maralinga tests had caused the deaths of Aboriginal people and that people were used for human experiments in underground bunkers at Maralinga: 'The cover-up, fraud, deceiving, etc., of the finding of deceased aborigines on the Maralinga/Emu testing range must not be allowed to continue any longer'.[41] At the Royal Commission he also claimed that 'mentally retarded people' were flown in by the RAF to Maralinga and placed into underground bunkers during atomic tests, a claim dismissed by Justice McClelland.[42] Separating the truth from the innocent misunderstandings, exaggerations and deliberate falsehoods has always been difficult in this shadowy tale. In truth, we do not know everything that happened in that warm spring of 1953, years before Maralinga saw its first mushroom cloud.

One compelling allegation was the 'deathbed confession' of David Bailey, in 2012. Bailey, a member of the RAAF construction crew that built the Emu base, confided in his nephew, Stephen Hogeveen. What he said is too ephemeral to prove, although it has the ring of truth. Bailey was a farm hand who enlisted in 1949 and requested discharge from service in 1956. He was an RAAF mechanical transport driver who was part of RAAF No. 5 Airfield Construction Squadron. During his seven years with the RAAF he was court martialled twice, although neither occasion was related to Emu (he stole ammunition in 1951, and in September 1952 he was absent without leave for a day.[43] For the first offence he was given 60 days detention, and for the second he was docked a day's pay). Upon discharge, his conduct during service was officially assessed as 'Very Good', and he was described as: 'An honest, unselfish type of airman with sound principles and good moral outlook whose conduct has been irreproachable but for whom an "exemplary" cannot be recorded either by reason of the 5 years' service qualification or because he does not quite measure up to the highest standard'.[44] His trade skills were deemed to be 'Satisfactory – The average tradesman, can be trusted to carry out his work with a minimum of supervision'.[45] He later ran a concreting business in civilian life, operating two trucks. 'He was a really nice guy and a hard worker', according to his nephew.[46] Bailey cared for his chronically ill first wife, Margaret, for many years until her death.

When Bailey wrote to the Department of Defence in 1997, requesting his service record, he added a postscript: 'I was involved in the 1953 October Atomic bomb test at Emu Claypans, S.A., but was not listed in records as such. 8 months Serial No. A21820 SACS. Original No. 169767 2 ACS, Woomera, 1949–1956'.[47] His official service record does not in fact show his time at Emu, although it does indicate he was at Woomera. This would be the usual practice, as RAAF involvement at Emu was initially kept secret, then later low key. One of Bailey's jobs at Emu was to help build the all-weather runway and the camp, using a large, tank-treaded bulldozer. He was also ordered to do sweeps of the bomb sites, looking for Aboriginal people to turn away before the tests began, although he never saw any. He told Hogeveen that Aboriginal people would have hidden from the RAAF men with their large and noisy vehicles. After the tests, Bailey received new orders. He told his nephew he was among a small party of RAAF men ordered to bury Aboriginal people who died as a result of one of the Totem blasts, most likely Totem I; Hogeveen suggests up to ten bodies may have been buried. Bailey told his nephew that he drove a large construction vehicle with the bodies loaded into the back. This story specifically alleges that these people were killed outright in the blast. According to Hogeveen,

> they were taken to a certain spot and Uncle David was told to dig a big hole with the bulldozer, which he did, and then they just pushed them out into the big hole and he was told to cover them up. And that whole crew was told that if they ever said anything to anybody, under the Official Secrets Act they would get 20 years in prison. So they were all very, very scared of what was going to happen to them.[48]

Bailey told Hogeveen when he revealed this story, 'look, this is really quite horrifying'.[49] As a justice of the peace for 30 years, Hogeveen had experienced this kind of thing before: 'I am sure he was giving me a dying deposition'.[50]

Hogeveen describes his Uncle David as a good man. A photo taken not long before his death shows him standing proud, wearing his medals, off to an Anzac Day service. He was a member of the Returned Services League and a strong Christian. Hogeveen believes that David, dying of a lung condition, was trying to clear his conscience before 'meeting his maker'.[51] He was a family man with a strong ethical sense who had no need to lie about an event decades earlier to his nephew, himself an upstanding citizen and an intensive care paramedic. Whether Bailey's deathbed confession is an accurate account of what happened, or another version of the more elaborate and fanciful story Toon told to the Royal Commission, is not at all clear. Bailey's story was not aired at the Royal Commission, since it was only told many years later.

No records have been found to support David Bailey's story, so it is not accepted here. However, circumstantial evidence means that it can't be wholly rejected either. The historian

Heather Goodall, who dealt with a substantial amount of evidence from Anangu people in the lead-up to the Royal Commission, and lived in Anangu lands for several years, did not hear of Aboriginal people being killed outright by the Totem blast (and she was well connected enough to have heard news of this magnitude). But she believes that the freedom of movement enjoyed by Aboriginal people in the area means that this story is plausible:

> The place was so big and so arid and grazing has been such a marginal or even failed activity, that there was a sense that people could move very effectively. That is certainly how people talked about what they had done in the past. So it is inconceivable that people weren't moving around that area.[52]

Ernest Titterton's work at Emu proved that anyone within just over 300 metres of the device as it detonated would be killed immediately. The presence of Aboriginal people within range of the early morning blast is entirely possible. As Goodall says:

> They would have been moving in areas related to water, because you can't travel without water and there was no effective way of carrying it, so you moved from waterhole to waterhole and you knew where the waterholes were. That means that there was some sense of a regular set of routes.[53]

Counsel of the Royal Commission Peter McClellan has also written about the complexities of oral evidence, most particularly distinguishing between 'real truth' and 'perceived truth'.[54] In evidence taken in Brisbane in October 1984, one witness told a story so outlandish that McClellan was embarrassed it was even discussed in the proceedings. Mark Earner had been a corporal in the Air Trials Unit based at Woomera, and was told that he would be working on a top-secret project.[55] He was a member of a team of six fellow RAAF workers at Woomera in a part of the armaments area, with special guards and a high fence. Stores arrived in crates, escorted on trucks from Port Pirie or Adelaide. 'We assumed, seeing that we knew the British scientists were around the place, this is what we were going to do. And we assumed that there were atomic bombs, but nobody told us specifically that they were.'[56] Earner believed that he and his team were assembling atomic bombs, and specified that he had been involved in assembling 11 altogether, two of which were not used. Under questioning, Earner said he believed he saw the inner workings of the atomic bombs being inserted. He claimed that the unused bombs were taken into the desert and burned, something that he found difficult and distressing.[57]

The Royal Commission concluded that there was no reason to disbelieve Earner's statement 'about his burning of what he believed were bombs at Woomera, but it is clear that whatever was burnt contained no radioactive material'.[58] Many years later, Peter McLellan was amazed when researchers attached to the Royal Commission told him Earner's story had some elements of truth:

> At Woomera they had been carrying out a major testing program on the casing design for atomic and later hydrogen bombs. There were many designs which had been mocked up in magnesium. In fact [Earner] had been sent to destroy casings which were being discarded and which the British wanted to keep secret from the Russians.[59]

The decades of anxiety generated by the tests affected service personnel as well as the Aboriginal people of central Australia. The witness, Mark Earner, had carried a secret and honestly believed he was complicit in dangerous activities. Peter McClellan said, 'His delivery was emotional and at times confused. He certainly did not look me or Justice Jim McClelland in the eye.'[60] Nevertheless, he had told the truth as he perceived it, and had little knowledge about what he had been doing. Such are the very human reactions to overwhelming secrecy and lies, and the threats made to maintain them.

Chapter 7
FLYING THROUGH THE CLOUDS

> The cloud, as we drew nearer, looked distinctly nasty.
>
> Wing Commander Geoffrey Dhenin, RAF, on seeing the Totem I cloud from the cockpit of his Canberra aircraft, just before flying into it.

> If you want to watch this thing, I advise you not to. It you have to look at it, pick which eye you want to lose.
>
> Advice given by Wing Commander Rose to Assistant Squadron Gunnery Leader Roy Cosgrove flying in an RAAF Lincoln towards ground zero as Totem I was about to explode.

The clue is in the name. Military aircraft flew through the mushroom cloud at Emu under the codename Operation Hot Box. Given 'hot' is a colloquial term for radioactive, the British propensity for whimsical codenames has a darker undertone here, with a name that seems to acknowledge that pilots and ground crew would be exposed to radioactivity. And indeed, the British aircrew flying their Canberra through the Totem I cloud received such a dose of radiation that sorties into the Totem II cloud were drastically scaled back. The Canberra was grounded for the second blast. The British atomic authorities were not known for their commitment to safety, but even for them the risks were too great.

Both the RAF and the RAAF contributed aircraft and crew to Operation Totem. The British had brought over the Canberra aircraft specifically for cloud-sampling duties. The RAAF contributed Lincolns and Dakotas, and dozens of air and ground crew.[1] The United States Air Force joined in too, flying their two B29 Superfortresses out of Richmond RAAF base, doing cloud sampling and showing up the deficiencies of the British safety measures. Whatever safety issues the Americans might have had in their own nuclear test program, they were noticeably stricter in their safety measures than the British. The Royal Commission found that

> [g]iven that the B29s were well sealed, that the crew used oxygen from the time that cloud contact was made until the time the aircraft landed at Richmond, and that eating and smoking were forbidden from the time of contact until after the crew had washed and changed at Richmond, the USAF [United States Air Force] crew members were placed at far less risk than members of the RAAF.[2]

Operation Totem was focused on the sky. A wide variety of aircraft played a role and there was much enthusiasm for airborne activities at Totem among the test authorities. Even a lesser known form of flying machine was slated to take part, an early kind of unmanned military drone called the Jindivik, co-created by British and Australian engineers at the rocket range at Woomera.[3] Early planning detailed in the Totem Panel minutes show an intention to use Jindivik flown from the Dingo claypan rather than piloted aircraft, noting

> it was not desirable that this target aircraft, which would become highly contaminated, should operate from Emu, and the only alternative seemed to be a separate operation based on Dingo claypan. Although this nominally would be a safe area, it might be necessary for a small aircraft such as the Prince to stand by to evacuate personnel in the event of likely contamination from the cloud. The whole proposal was still being discussed in [the] U.K., and in the meantime the Minister's approval had not yet been sought for the use of the aircraft.[4]

In the end, despite Woomera technicians working overtime, the automated Jindiviks were not ready. Instead, pilots and ground crew risked their lives, and many paid with their future health.

While sampling the atomic cloud from the air seems a logical (if dangerous) way to obtain data on these particular kinds of explosions, these air-sampling sorties had another purpose. According to Penney's evidence at the Royal Commission, flying military aircraft through the Totem clouds was connected with developing and refining techniques to monitor the nature of US and Russian weapons. Counsel assisting the Royal Commission, Peter McClellan, asked Penney about this: 'one of – and perhaps the main – reasons for the cloud-sampling exercises was to inform the United Kingdom about methods of analysing the Russians' or the Americans' explosion, or the French or whatever?'[5] Penney agreed this was correct and said the security services had requested these experiments to create a 'library' of data that could be matched with the atomic clouds of other countries. The clouds could be sampled a long way from where the bombs were detonated, allowing deductions to be made about nuclear testing activities. Penney went on, 'I am not telling you any secrets here, because this was all discussed with the Russians at the Test Ban Treaty [in 1963], when we were saying, "Let's make a control system"'.[6] McClellan speculated that the Russians would have been able to establish the yield from Mosaic G2 at Monte Bello in 1956, the largest atomic weapon tested in Australia. Penney wasn't sure:

> I do not know whether the Russians could get near enough to get any samples. You could get samples around New Zealand, but I do not know whether the Russians had ships or aircraft around New Zealand. But suppose they did; what they got was a ratio [of] plutonium to burnt plutonium ... So unless you have got a library of these things and look for finer differences, you may not be able to get the yield.[7]

Britain may have started from an alarmingly low base, but every atomic test held in Australia had associated experiments that increased their knowledge, including their ability to measure weapon yield.

Prior to their atomic testing program, the British test authorities had made erroneous predictions about the likelihood of aircrew contamination. This meant that RAF personnel flying into the Operation Hurricane cloud in 1952 had no radiation monitoring gear in their cockpits, an almost unbelievable dereliction of safety duty.[8] Having learned their lesson from Hurricane, some monitors were installed into the aircraft used at Totem allowing the authorities to track the readings of radiation following aircraft entry into the fallout plumes. High readings were recorded in all aircraft that came close to, or entered, the atomic clouds for Totem I, and consequently aircraft did not sample the Totem II cloud directly after detonation.[9] Two RAAF Lincolns did enter the Totem II cloud about 643 kilometres from the blast site, and samples were taken about 10 hours after the detonation. On arrival back at Woomera, the crew and both aircraft were found to be radioactively contaminated.[10] The responsibility for decontaminating aircraft, shared jointly between British and Australian personnel, was not taken sufficiently seriously, according to the Kerr report: 'The Committee found that information on the hazard faced by decontamination crews was most inadequate'.[11] The Kerr report preceded the Royal Commission by a couple of months and pointed to serious problems that Jim McClelland would take on board to further investigate. It is likely that ground crew were at even greater risk of radiation exposure because they came into contact with the dust and other debris from the returning aircraft every day, the stuff getting into their clothes and onto their skin. They were never warned of the risk.

The Royal Commission examined the dangers faced by RAAF workers in detail. During evidence, Peter McClellan asked Rex Naggs, a member of the 82 Wing RAAF engine maintenance crew at Woomera, about the risk into which he was placed as a teenager:

> McClellan: You say [an English officer] checked out the planes and discovered that they were much more radioactive than had been predicted. He checked your overalls and your body and said that both were radioactive, and you were ordered into the showers. He would not let anybody touch your overalls, and you saw them removed on a 6 foot stick after you had cleaned out the pockets. Did that concern you?

Naggs: It started to worry me then, yes, that he was scared of my overalls and I had been wearing them.

McClellan: For two days?

Naggs: Well, I do not know exactly how long I had been wearing them because – I was a 19 year old, it was all brand bloody new, it was the first operation I had ever been on, and it was a bit peculiar when they – he would not go near those overalls.[12]

Naggs' contaminated overalls were buried, as were those worn by his fellow workers. Since Totem, he had been told by the RAAF to have blood tests regularly, for years, without being given a reason or seeing any results. However, in January 1984, he was diagnosed with leukaemia, and had been undergoing chemotherapy when he gave evidence at the Royal Commission.[13] He did not know whether his service at Woomera, or his later service in the Vietnam War, was implicated.

The RAAF contributed 10 Lincoln bombers based at Woomera and Richmond for air sampling and two Dakotas based at Woomera for ground contamination surveys as well as air and ground crew support.[14] Group Captain David Colquhoun, commander of Amberley base near Brisbane, led the RAAF effort from Woomera, and Group Captain Maxwell Watson was the RAAF representative on the Totem Panel.[15] Colquhoun commanded RAAF personnel to undertake a variety of tasks associated with Totem, including Task Totem Alpha, an aerial survey using three Woomera-based Dakotas to determine possible ground contamination caused by fallout from either Totem blast.[16] The surveys were guided by AWRE's

Charles Adams, who was based at Emu as scientific supervisor for Totem. RAAF Lincolns carried out additional operations, titled Totem Beta (to gather airborne radioactive dust) and Totem Gamma (air support and meteorological activities).

Five Lincolns based out of Woomera flew through the Totem I cloud to undertake air sampling.[17] These aircraft made a total of 15 traverses of the Totem I cloud, and therefore had 30 entry and exit points that would enable calculation of the dimensions of the cloud.[18] The captain of one of the Lincolns radioed to the tower at Woomera that radiation readings on the aircraft were worryingly high and that once the aircraft landed no-one should approach the planes.[19] Crews landing at Woomera after flying into the cloud were treated to the 'water buffalo'. A vehicle would meet them on the tarmac and shower the aircraft and the crew with water, often too cold for the liking of the air crew, and almost certainly insufficient to deal with the radioactive material that had accumulated on both the crew and the fuselage.[20] Adams provided the AWRE with the top-secret scientific data gathered from these operations.[21] The Royal Commission found, '[i]t was negligent to allow aircrew to fly through the Totem 1 cloud without proper instructions and without protective clothing'.[22]

Another RAAF venture was Operation Dead Heart (also an unfortunate code name), in which an 82 Wing Lincoln was used to obtain photographic data from the Totem tests, to augment information recorded by the camera towers on the ground. Confusion over whether or not these photographs could be given to the media caused some consternation at high levels within the RAAF. Charles Adams initially advised the RAAF that because the photos could not reveal the aircraft's height and distance from the explosion, they could be cleared as unclassified. However, after subsequent discussions with the air officer commanding Home Command the photos were withdrawn from public view and declared secret, since 'information revealing that the RAAF had participated in Operation "Totem" was not to be released'.[23] This ruling was then reversed because an air force press release had already discussed aspects of the RAAF's role in Totem.[24] As was often the case in the atomic tests, right hands and left hands were at odds with each other.

Astonishingly, the Australian authorities had not foreseen the possibility of contamination to RAAF aircraft and crew. They were even further behind in their forecasts than the British. The RAAF command was completely taken by surprise when aircraft came back hot from their missions and had to implement measures on the fly. The Royal Commission quoted an official Operation Totem report referring to the RAAF Lincolns involved in cloud sampling as saying 'assurance had been given by the Scientific Director, Emu Field [Adams] that no such problem was likely to exist as far as Totem Beta aircraft were concerned'.[25] Group Captain Colquhoun related an anecdote at the Royal Commission about a group of RAAF ground crew gathered in a hangar at Woomera where a doctor ran a Geiger counter over them. On one man

> as it got to his hip the Geiger gave a very strong number of counts and we stopped, investigated and the lad had a rag in his hip pocket which he then said he had wiped grease off the union between the wing and the fuselage and this was heavily contaminated.[26]

The fact that no-one seemed to have foreseen this as a problem speaks to the general neglect of safety at Totem.

The Royal Commission also found that RAAF personnel

were not included in radiological safety orders for Emu Field, so no health control procedures were in place at either Woomera or Richmond for RAAF personnel, whether flight crew or ground crew.[27] The risk to pilots and ground crew from the Totem clouds was 40 times higher than the highest levels measured in the first five flights after Operation Hurricane, because the Totem flights were taken much earlier after detonation compared with those at Hurricane, which only gathered samples 24 hours or more after detonation.[28] Colquhoun wrote to the senior air staff officer at Home Command Headquarters on 21 October, saying, 'All the scientists are somewhat nonplussed with the amount of radioactivity registered in Totem I and are most apologetic that the matter was not fully covered in the briefings, and the issuing of equipment before Totem I'.[29] He added rather archly, '[t]hey are co-operating fully now'.[30] Nevertheless, Colquhoun wasn't overly worried:

> I would like to stress that there is no risk or undue concern over this radiation business. I am assured that the risk to aircrew is negligible but there is a possibility that the ground crew could, by swallowing radioactive particles, suffer some discomfort in later years. The latter is the reason for our precautionary measures.[31]

Ground crew, often shirtless, worked for hours cleaning and maintaining the aircraft.

Part of the extensive testimony about the risks to RAAF personnel given at the Royal Commission was that of retired squadron leader Lancelot Edwards, who gave evidence in Brisbane. He had been a wireless operator aboard one of the Woomera-based Lincolns, and his vivid account was featured in media reports. 'After the second mission [to sample the Totem II cloud], Mr Edwards showered, without soap, 14 times before his radiation levels returned to normal', reported the *Canberra Times*.[32] In his statement to the Royal Commission, Edwards said:

> Our aircraft became contaminated during Totem I and received further doses in Totem II. To my knowledge no decontamination procedures were carried out on the aircraft until after returning to Amberley some time later. In late December, 1953, all crew members were required to attend the Medical Section at Amberley for blood testing, presumably to check the results of radiation exposure. I was not advised of the results of those tests.[33]

In 1959, Edwards was diagnosed with cancer of the thyroid, a condition that was accepted as being caused by his radiation exposure. He was treated for the cancer and in 1981 he received administrative compensation from the Commonwealth Government.[34]

The Lincoln bombers were all contaminated after their operations at Emu, and were subjected to decontamination processes both at Richmond base near Sydney and at Amberley near Brisbane.[35] Intervention by the American aircrew at Richmond meant that the aircraft received a higher level of cleaning before travelling onward to Amberley than they might otherwise have done. According to an AWRE report, Lincolns flew through the clouds of both Totem blasts, unlike the British Canberra aircraft, which was held back from Totem II:

> The 'Lincolns' operating from Richmond (N.S.W.) also became contaminated, one seriously, but full precautions

were taken there, the RAAF being instructed in these matters by the United States Air Force, who also loaned a monitoring instrument of the 1021 type. These planes were also flown to Amberley for decontamination.[36]

At Amberley, nine aircraft were found to be contaminated, of which three were 'highly contaminated' and two were just above the tolerance levels.[37] The hottest aircraft was moved to a special site for a thorough systematic decontamination process intended to make the plane serviceable again. The report recommended that future aircrew be carefully monitored for the effects of radiation and that: 'If aircraft participate in future trials consideration might be given to means for reducing their times in the clouds and thus lessening the problem [of difficult-to-remove contamination]'.[38]

In planning for the airborne aspects of Totem, the British specifically chose the Canberra for entry into the cloud immediately upon detonation because of its suitability for air sampling. The Canberra was roomy and fast, and it could remain stable at high altitude when equipped with a quiver-full of external measuring devices. On 15 October the Canberra flew through the cloud of Totem I, just six minutes after the device had been detonated. The operation was designed to obtain specific information requested by the RAF.[39] While at Farnborough in the United Kingdom, the aircraft had been fitted with special equipment designed by the Electronics Division of the Atomic Energy Research Establishment at Harwell. It was also fitted with special filters at the request of the Radio Chemistry Division of the AWRE.[40] Official AWRE footage of the Canberra flying into the cloud looks chilling, and indeed moving to anyone with any empathy. The craft looks so small in the huge skies of Emu Field, with the atomic cloud looming in front of the three largely unprotected British RAF crew inside.[41]

The Canberra aircraft had left Marham in Norfolk on a Sunday morning, 13 September 1953, and flew in a series of stages, arriving at Woomera on 19 September. The crew were Wing Commander Geoffrey Dhenin, who piloted the plane, the navigator Wing Commander EW Anderson, and radiological observer Group Captain Denis Wilson from the RAF Central Medical Establishment. Wilson had a unique background, in that he had been a consultant radiologist for the RAF, attached to Harwell, and had recommended aircraft-based radiological experiments as part of the atomic testing program.[42] The Canberra crew had visited X200 before Totem to inspect the claypan landing strip, declaring that it was suitable for emergency use only. They based themselves at Woomera and participated in rehearsals and the actual operation from there. Dhenin had explicit, yet impossible, orders to 'ensure that neither the aircraft nor ground personnel are subjected to radiation hazards either during the trial or as a result of the contamination of the Canberra aircraft'.[43] Since the mission was specifically tasked with assessing the risk to air crew from radioactivity during passage through the atomic cloud and in its vicinity, there is no doubt that there was going to be significant contamination.[44] Dhenin and crew were also required to gauge whether pilots would be dazzled by the initial flash of the explosion and to determine how an aircraft would behave inside an atomic cloud. As Dhenin later reflected, '[a]part from the normal danger of looking at atomic explosions, a blind pilot cannot regain control of an aircraft – and a Canberra once out of control hits the ground very fast indeed'.[45] The crew were exceptionally brave.

On the day of Totem I, Dhenin's aircraft took off from

Woomera at 5.50 am. It was accompanied by an RAAF Lincoln from 82 Wing that acted as a 'radio link' aircraft that would circle around about 100 kilometres away from the explosion and provide a back-up radio link should the Canberra aircraft's communications be taken out by the atomic blast. The Lincoln also carried stores and rescue equipment in case Wing Commander Dhenin's aircraft was forced down. The Lincoln and the Canberra maintained radio links throughout the Totem I sortie.

When Dhenin's Canberra arrived at X200, it circled north of the test site at 35 000 feet until the test countdown began, then it began to descend. As the bomb exploded, Dhenin turned his plane at an angle of about 30 degrees to the burst. At that point they were 16 kilometres away at 20 000 feet altitude. According to Group Captain Wilson, 'The flash lit up the cockpit and we were immediately able to watch the remainder of the explosion'.[46] Six minutes after the flash, the plane entered the 1.8-kilometre-wide cloud tangentially at 14 000 feet and stayed inside it for over nine seconds. Instruments in the aircraft started to register radiation about 457 metres away from the cloud, and while the crew sat in their hot box aircraft for those nine seconds, tossed around by the extreme turbulence, the external instruments registered approximately 2000 roentgens per hour, fortunately dropping sharply soon after. According to radiation scientist Geoff Williams, with a dose of that size, 'even a very short exposure (minutes) would have massive deterministic health effects on the human body'.[47] The crew wore oxygen masks and the cockpit was sealed against possible leaks 'by fabric, dope [adhesive] and selotape'.[48] An AWRE report produced in 1954 noted that the RAF Canberra crew described the atomic cloud as similar to 'a dense London fog', requiring that the aircraft's lights be turned on while inside it.[49] London fogs were never so dangerous.

Once they emerged from the cloud, they made runs above and below it as it rose, roiling and rushing, to the upper atmosphere. 'Dosage rates of 300 [roentgen] per hour were noted above the cloud and 400r below some two minutes later.'[50] The cockpit instruments were still recording radiation, in reducing quantities, during the flight back to Woomera. A second Canberra aircraft joined the final stages of the flight to Woomera, just in case Wing Commander Dhenin's aircraft had lost its air speed indicator. All was well, though. The plane landed normally and taxied to the end of the runway. The crew were immediately showered and ushered into clean clothing. A Geiger counter was run over them, after which they were allowed to go to breakfast. According to Wilson, '[t]here was no contamination in the cockpit as far as could be ascertained, our clothing becoming contaminated on the way out of the aircraft.'[51] This comment points to an issue that was not fully appreciated at the time – that ground crew exposed to the outside of the aircraft were potentially at greater risk than flight crew.

After the plane had cooled down, its contaminated areas were charted. The machine was sprayed with a detergent-based decontamination solution from buckets, using a stirrup pump, and scrubbed with long-handled brushes. The one-hour cleaning operation was completed using a high-pressure jet of water. After it had been left standing on the runway for 48 hours, it was surveyed again. Heavy contamination was found on the leading edges of the wings, on the Perspex nose and on the engine coverings. The quartz fibre dosimeters worn by the crew were analysed, and some readings were 'off the scale'.[52] The film badges worn by the crew showed less radiation, for reasons that were unclear. Wilson commented:

> These film badge dosages brought us within 0.5r of the maximum permissible dosages allowed for the operation by the Minister of Supply, who stated that since all the RAF requirements had been fulfilled on the first sortie, that they did not feel justified in asking us to collect samples solely for them in the second explosion.[53]

The crew wanted to fly through the second cloud, but obeyed their orders and watched it on the ground, some 6.5 kilometres from ground zero. No matter – their sampling had been successful and they had retrieved usable data for the scientists and RAF brass back in the United Kingdom.[54] For the second blast, rockets were used – unsuccessfully – to attempt to obtain samples from the cloud. As Denis Wilson wrote, 'It was interesting to note that through an alteration in the surface wind the sample rockets did not connect with the cloud – thus emphasising the value of aircraft sampling and in this case showing how useful it might have been had we been airborne, standing-by, at the time'.[55]

The Canberra crew flew their plane back to Britain, leaving Woomera on 3 November and arriving home three days later. There they were subjected to a battery of medical tests that did not find anything untoward. The engines of the plane were another matter. Wilson estimated that about half a tonne of contaminated air had flowed through the two engines, followed by thousands of tonnes of non-contaminated air on the flight back to the United Kingdom. Even so, and despite the cleaning operations at Woomera, the insides of the engines were found to be 'visibly contaminated with radio-active dust which must from its activity have come from the inside of the cloud'.[56] When Wing Commander Dhenin was consulted he found further contamination in other parts of the port engine, which was sent to the manufacturer, Rolls-Royce, to be cleaned and rebuilt.

A year after Totem, Denis Wilson speculated on the hazards posed to commercial airliners that might encounter the radioactive particles from a growing number of nuclear tests around the world. He co-authored an article with Geoffrey Dhenin published in *Proceedings of the Royal Society of Medicine*; Wilson contributed the first half and Dhenin the second. Many insiders already knew that Earth's atmosphere was becoming increasingly contaminated by the huge surge in testing. Wilson picked up on this:

> It is theoretically possible that an aircraft flying on a standard course at an altitude around 45 000 feet for some hours might pick up and push into its pressurized cabin (and its passengers) measurable amounts of fission products. The medico-legal implications of a jet airliner full of rich and influential business men getting in such a position are full of intriguing possibilities![57]

Wilson was not overly worried, given that passenger aircraft in 1954 did not fly that high, and if they did, the radioactive particles could be filtered out, though he noted that currently 'no passenger aircraft carries these filters'.[58] However, despite the lighthearted tone of some of this article, it clearly indicates that the crew who flew through the Totem I cloud did have cause for concern. In Dhenin's section of the article he specifically addressed his experience in the Canberra, listing five risks that he and his crew faced:

1. Hitting something solid in the cloud.
2. Engine 'flameout' due to the dust.

3. Losing the airspeed indicator due to dust.
4. Inhaling or swallowing radioactive dust.
5. External radiation.[59]

The first three risks were significant but manageable, he determined. The Canberra was equipped with ejector seats and the crew trained in their operation. He was perfectly prepared to carry out 'a good belly landing'[60] on country that was relatively flat if he lost the engines, and he had RAAF backup to call out his airspeed if needed. He then considered the risk of radioactivity entering the cockpit and the whole aircraft being overwhelmed by radiation:

> The fourth risk we countered by sealing the cabin – cutting off the pressurization air from the engines and sealing all possible leaks by fabric, dope and selotape. Though the cabin environment was scarcely enhanced by this treatment, we were, of course, using oxygen and preferred the discomforts of low pressure and no ventilation to the hazards of cabin contamination. The fifth risk we could do little about except to calculate what radiation dose we could accept, and arrange to enter the cloud at a time appropriate to the dose.[61]

In a sense, the RAF Canberra crew were able to provide more informed consent than any RAAF crew, because they had Wilson's expertise on board.

An RAAF Lincoln from Richmond that tracked the Totem I cloud to Townsville, two days after the explosion, lost an engine while returning along the Queensland coast and had to make an emergency landing at Amberley. Signaller, later Squadron Leader, Morris Newman was onboard. In a statement to the Royal Commission, he said that the crew had radioed ahead so that the ground crew would be ready for their unscheduled and urgent arrival. 'Upon landing at Amberley, there were no special precautions taken for the aircraft and we simply parked in the usual dispersal area.'[62] When Newman and the rest of the crew were taken back to Richmond, there were no requirements for special showers or removal of their contaminated clothing or flying gear.[63] He was aged 53 when he supplied this statement, and had already experienced significant health challenges, including eye and back problems. A decade later he developed pancreatic cancer and died at 69.[64] In the years leading up to his death he was – after a protracted argument and government stonewalling – provided with some Commonwealth assistance for the medical expenses related to his cancer.[65] His daughter, Michele Newman, remains angry at the neglect and recklessness that left her father vulnerable to harm.[66]

The planes assigned to sample from the cloud were not the only ones affected. Charles Geschke was the captain of an Australian Lincoln bomber during Totem. In his statement to the Royal Commission, he told of taking 'clean containers' on his aircraft from Richmond to Townsville. He had been instructed to avoid the Totem I cloud on his journey:

> Whilst on route, however, I noticed that the radar had produced some peculiar flashes and I decided to turn on the radiation detection instruments. Those instruments revealed a high level of radiation. In fact they reached maximum deflections. I then flew the plane through prescribed pattern flying in order to determine the precise position of the cloud. It was clear that the information given by the previous aircraft as to the position of the

cloud was incorrect. On arriving at Townsville I reported the incident fully. There was no decontamination of any member of the crew or any checks done for radiation contamination.[67]

No-one seems to have foreseen that any aircraft flying from Sydney to Townsville could encounter the atomic cloud and become contaminated. A restricted teletype message sent on 28 October from Eastern Area Detachment Woomera to the commandant of 82 Wing, just after Totem II, said, 'secure room required to house quarantined flying clothing and equipment all crews and [passengers] … until 2 Nov'.[68] They were wearing and using items that were dangerous.

One official voice raised in alarm about the possible hazard to RAAF personnel was that of Air Vice Marshal EA (Ted) Daley, director general of Medical Services for the RAAF. In a secret minute paper, he briefly outlined his concerns, calling the information he had received about Totem 'disconcerting'.[69] He had been in touch with Group Captain Wilson from the RAF Canberra, and was aware of the huge cabin dosage received by the British crew. He said:

> before the Monte Bello work [Operation Hurricane], I firmly raised the question of exposure of R.A.A.F. personnel to this radiation … we were firmly told that this was not a hazard on that occasion. It does appear that there was a hazard then just as there has been a practical hazard this time [during Totem].[70]

He noted that all RAAF crew involved with the Lincolns at Totem had undergone medical testing, drily adding: 'It is fortunate that no apparent dangerous doses have been received by R.A.A.F. personnel, but this would appear to be no fault of the British Ministry of Supply'.[71]

Daley continued his veiled and unveiled criticism of the British Ministry of Supply, whom he mostly blamed for exposure of RAAF personnel to radioactivity. In a secret document that later formed part of a minute paper on radiological safety compiled by the RAAF, he mentioned the case of Flight Lieutenant Richard Nettley, an RAF officer who was on exchange to the RAAF, based at Amberley, and who had been exposed to radiation from one of the Lincoln aircraft used in Totem. He and the six other crew of the Lincoln had flown in and out of the Totem I cloud for three hours.[72] Daley said that his previous concerns had 'been justified but that information and assistance from the British Ministry of Supply has not been adequate'.[73]

The two United States Air Force B29 Superfortresses also became highly contaminated, although the Americans were more safety-conscious than the British and looked after their crew better.[74] The B29s were backed up by two C54 aircraft for transport support.[75] The Americans were to stay at least 645 kilometres from the blast, after flying out of Richmond. The US crew had sensitive monitoring equipment onboard, and they picked up intense radiation late in the evening of the first blast.[76] They located not the visible atomic cloud, but unseen small patches of radiation. According to Symonds, '[t]heir sensitive radiation detection equipment had reached its full scale reading and remained so until the contamination on the aircraft had decayed to a certain degree with time'.[77] The crew reported that the activity they detected was 'the most intense they had ever encountered in their experience'.[78] The American crew insisted that their aircraft be isolated for 72 hours after returning to Richmond. Their concerns were noted by the British and Australian crew: 'it is clear that the Richmond-based detachment was fortunate that the experienced USAF [United States Air Force] staff were available to advise on the degree of contamination of aircraft and personnel'.[79]

RAAF pilot Allen Clark, who retired from the air force in 1968, stated in a submission to the Royal Commission that he had flown one of the Lincolns that entered the Totem I and II clouds, and it remained in service despite receiving doses of radiation. 'At no time during my entire flying career in the RAAF was I ever advised that any aircraft I was flying in might be radioactive and I was never given any protection. I never saw aircraft or personnel checked for radiation.'[80] Like the Aboriginal people, many of the surviving service personnel have carried anxiety for decades from the time of the atomic tests.

RAAF gunnery leader Roy Cosgrove gave a vivid account in his submission to the Royal Commission of flying towards Totem I to capture pictures from his Lincoln. He was under orders to take photos of Totem I from an altitude of 8000 feet, commencing from a distance of 40 kilometres from the explosion:

> At the moment of the explosion the aircraft was engulfed in a very intense white light similar to the beam of an ophthalmoscope being shone into one's eyes. It was an eerie sensation. Being a Christian, it is what I imagine the light of God being like.[81]

Cosgrove also gave a small glimpse of the attitudes that have persisted to this day when it comes to compensating and supporting service personnel who participated in the atomic tests in Australia. Not long before the Royal Commission, he had watched an episode of the current affairs television show *Sixty Minutes* in which Sir Howard Beale, once the minister in charge of the Australian end of the test series from a comfortable distance in Canberra, answered a question about the allegations of harm that were then emerging from atomic veterans. Cosgrove remembered that Beale said: '"All servicemen are prone to exaggeration". Whilst they [sic] might be true, the manner in which it was said really annoyed me. All of us gave faithful service to our country and we deserve better'.[82]

The Royal Commission found that, rather than being prone to exaggeration, service personnel were probably the opposite, given the mountain of evidence about the lax safety for all military personnel, especially those from the RAAF. Here are some of its findings:

> Aircrew of the Lincoln aircraft at Totem 1 should have been supplied with radiation monitoring devices and given instructions as to the behaviour of these devices when in the cloud or a contaminated aircraft. The failure to provide this equipment and instructions was negligent. Ground crew should have been similarly equipped and instructed.
>
> Air and ground crew of Lincoln aircraft used for Totem 1 suffered exposure to radiation but the doses which they received are now impossible to determine accurately. It is unlikely that the doses exceeded the level of dose which others in the program were authorised to receive ...

> [T]here was lack of foresight shown in the failure to institute a proper system of decontamination for RAAF aircraft at Woomera before the Totem 1 detonation.[83]

John Clarke QC conducted a review of veterans' entitlements in 2003 that reaffirmed the Royal Commission's findings that air crew involved with Operation Totem were not properly protected. Clarke's report stated:

> It was negligent to allow aircrew to fly through the Totem 1 cloud without proper instructions and without protective clothing. As in Operation Hurricane, ground crew and aircrew of the Lincoln aircraft at Totem 1 should have been supplied with radiation monitoring devices and instructed in their use, and failure to do this was negligent.[84]

In a significant pronouncement on risks faced by Australian military personnel involved in British atomic tests, Clarke found that:

> Apart from involvement in wars, other conflicts and overseas deployments, it is difficult to conceive of another Australian military operation in the 20th century comparable to the [atomic] tests' scale and risk of harm to individuals … There has been an inadequate response by successive governments over many decades. It is a sad fact that the recognition of the unusual hazards faced by the participants has not led to prompt action to ensure a more appropriate compensation arrangement with ready access, given the nature of the hazards.[85]

Another 14 years would pass until those affected by the tests would receive assistance through the Commonwealth-funded 'Gold Card', which provides unlimited medical treatment. In 2017, federal parliament passed legislation to extend the Gold Card to those harmed by all atomic tests on Australian soil, and the 2017–18 budget finally made provision for all Australian participants in the British nuclear tests in Australia in the 1950s and 1960s, including Aboriginal and other civilians within the test areas, to receive the Gold Card.[86] By then, few of the Australian air and ground crew who flew in or tended the Operation Totem aircraft were still alive.

Chapter 8
THE PEOPLE'S WITNESS

> It stood over the desert like a grotesque tree and for a while, until the winds seized it and distorted it and bent it to a random shape, there was an unwilling, monstrous quality of grace about it.
>
> James Cameron, British journalist, describing Totem I's atomic cloud in the *Age*.

Operation Totem marked the first time the media was allowed in to a British atomic test site as part of the official party. Journalists had been excluded from Operation Hurricane at Monte Bello so found a site to witness it unofficially. They would have been excluded from Emu too, if they hadn't applied sustained pressure. The British authorities, led strongly by Frederick Lindemann, were wary of media interest in the Totem tests, lest it led to unhelpful public speculation. In a note to Winston Churchill in August 1953 Lindemann wrote about media presence at Emu:

> I do not like the idea because it will create an undesirable precedent and because I am anxious to play down the publicity as the first test is of a quasi-tactical bomb to meet

C.O.S. [Chiefs of Staff] requirements which will be, and is intended to be, a smaller explosion than at Monte Bello.[1]

He added ruefully: 'One of the difficulties is that the press are often admitted in Nevada on similar terms'.[2] Indeed, the United States (while keeping its central secrets) allowed greater media access to its nuclear tests. The British were instinctively repelled by any such thing and would have happily left the media to do nothing other than reproduce their media releases. 'Press to be carefully shepherded and denied access to scientific and other staff', Lindemann wrote to Churchill when discussing the media request for access to Totem I.[3]

Although the media was allowed to watch Totem I, they were given no information, hinting that the real motivation for having journalists present was propaganda. The Fourth Estate has a role as 'the people's witness', to act as an intermediary between the public and the other three estates of government, the executive, legislature and judiciary. But in the case of Totem I the media was given 'almost no indication of any hazard which might arise for the Australian population', as the Royal Commission put it.[4] This meant it could not undertake Fourth Estate duties.

Journalists were even further down the pecking order than the other 'tourists' who were not wanted on site. Ben Cockram, deputy high commissioner in Canberra, had written to Prime Minister Menzies presenting rebuttals to any case for media presence, including parallels with American journalists. He set out the grounds for refusing media presence:

> (1) The Americans no longer find it possible to admit press representatives to the tests of atomic weapons (e.g. the last island test as distinct from the Nevada tests).

(2) That the reason for the change was that the U.S. administration did not find it possible at the first Bikini test to prevent the publishing, by the press, of details of a high security rating because sufficient consideration could not be given in the short time available to what details should not be published.

(3) That the mounting of a test is an undertaking fraught with the possibility of mishap and that the presence of the press inevitably leads to an intense desire to meet some particular date. It has been rumoured that, at the first Bikini test, there was a failure to gain the maximum information from the test because the presence of pressmen led to a strong desire to meet a particular schedule which resulted in failure to mount all the measuring instruments before the test itself took place.[5]

The Australian Prime Minister's Department was also no fan of the press. In fact, its Assistant Secretary Allan McKnight told Cockram that, in the case of Hurricane, even though journalists were not part of the official party, '[t]he outstanding difficulty at our end arose with the Press'.[6] He suggested: 'It may be that the almost hysterical requests which were made by the Press for the right to attend at Monte Bello were caused by the fact that this was the first British atomic test; there may not be the same eagerness to attend the second test'.[7] That turned out not to the case.

Media spotlights and atomic weapons do not go well together in the eyes of the government. In 1952, after decades of intense lobbying by the UK government, a D-notice system, modelled on a long-standing British system of the same name, was established in Australia in the lead-up to

Hurricane. D-notices were secret, voluntary agreements by the major Australian media organisations not to publish certain information on the grounds that it could compromise national security. McKnight complained that a lot of departmental time was taken up setting in place the D-notice restrictions about what the media could and couldn't publish. 'I would, therefore, suggest that we try quite early in the piece to have these principles settled.'[8] It must be said, however, that media coverage of the Totem tests constrained not just by D-notice, but also by the media's willingness to behave like patriotic cheerleaders rather than scrutineers. Investigative journalists looking for the truth about the Totem tests were nowhere to be seen.

Media management for Operation Totem was carefully planned, and extremely successful, in that the official line prevailed in both the British and the Australian press. A meeting of the Totem Panel in July 1953 approved the first Australian press handouts for the tests, which emphasised the calibre of the Australian military personnel who had established the site. It was a clever redirection away from weapon information towards popular pride in Australian service personnel:

> Unknown even to their closest relatives, members of a special [*sic*] picked work force from the Army and R.A.A.F. have been working up to 70 hours a week to prepare the atomic bomb site deep in the Southern Australian desert. To narrow the security risk and minimise the terrific supply of problems, every man chosen for the operation had to be expert at two or more jobs.[9]

The handout made clear that the test site was completely secret, specifying only that it was 'somewhere in the

80 000 square mile prohibited area which includes the Long Range Weapons Establishment at Woomera'.[10] The statement lauded the tight security that enabled a large infrastructure project like Emu Field to proceed without any publicity before the official announcement.

The Australian Totem Panel under chairman Jack Stevens considered media arrangements for the trials among its many duties as detailed in the secret minutes of the panel's meeting of 24 June 1953:

> The Australian attitude was that facilities were being created for U.K., and that whatever U.K. desired regarding the presence of the press, would apply. It was proposed to release from time to time background information in Australia, which the press could publish, and they were also permitted to publish observable facts. Background stories would be teleprinted to London before release in Australia, and these would, of course, only deal with unclassified aspects of the test.[11]

The channelling of media attention into safe byways deflected difficult questions about atomic testing in places people lived.

A remarkably obedient and loyal media in both Britain and Australia took up the cause with alacrity, without insider knowledge or investigative verve. An editorial in the Sydney *Sunday Telegraph* in August 1953, soon after the Emu Field tests were made public, declared that the tests would 'decide the type of atomic weapon that Britain will develop'.[12] The author, Harold Dvoretsky, wrote that atomic materials were costly, and weapons development was competing with industrial uses of uranium. 'This factor has prompted the British Army to make a new study to decide which weapon is best to adopt.'[13] This was not quite correct, as the decisions about atomic weaponry were taken by the inner sanctum of the British government, in consultation with the weapons scientists, particularly William Penney and his small band of experts, rather than the army. The military input was generally concerned with target response studies intended to shed light on the effect of atomic weapons, to understand how an atomic war would play out. The *Telegraph* author confidently asserted that troops on the ground after an atomic bomb drop would be unaffected:

> atomic weapons for the most part, whether used defensively or offensively in ground warfare, will be exploded above the ground. Not only does this add more devastating effect, covering, like shrapnel, a greater area, but it also leaves ground under the bomb explosion free from radiation effect. If this radioactivity lands, it may lead to new problems on when to use the weapon. But for the most part the radioactivity will be lifted into the atmosphere and carried away and will lose potency before it can land and do any damage. This means that troops can pass over an area where the atomic weapon has been dropped within a few hours without harmful effects.[14]

The level of misunderstanding in this passage is extraordinary, even for the time.

Much Australian media coverage highlighted and celebrated Australia's patriotic motives for hosting the tests. The Melbourne *Herald* on 3 August 1953 hailed the role of the Woomera range:

> Woomera is already an Anglo–Australian defence asset. Its installations are only specks in a 1500 mile corridor of

saltbush and desert. But when its tests bring to bear new processes that can be applied to the big uranium deposits in South Australia and the Northern Territory, the effects may be felt throughout our national life.[15]

The story misleadingly stated that 'Australian scientists will be working with Sir William Penney and other atomic researchers from Britain', when in fact most Australian scientists were denied any scientific information at all about the tests, with the exception of Ernest Titterton.

Indeed, the considerable British anxiety about control over media information is echoed in a series of documents concerning Leslie Martin's role in both Hurricane and Totem. Martin was at that time a defence science adviser as well as a physics academic at the University of Melbourne. The British High Commission had requested complete secrecy about his attendance at the Monte Bello test and his involvement in Totem. The Australian correspondence on this issue seems to push back at Britain's reluctance to give Martin access to some information about the tests. Bureaucrat Allan McKnight was supportive, while agreeing to keep Martin out of the spotlight. In a letter to George Davey at the UK High Commission in Canberra, McKnight said: 'The Australian authorities agree that no public announcement should be made prior to the [Totem] tests regarding Professor Martin's attendance. If some statement becomes necessary some cover plan might be devised.'[16] McKnight also provided the assurances required by the British high commissioner about Martin's security rating, pointing out that his role as a defence adviser meant he was signatory to the provisions of the Australian Crimes Act, 'which corresponds, in part, to the British Official Secrets Act'.[17] Despite this, the British deemed it necessary to keep his direct involvement secret until his role was clarified in 1955 by his appointment as chairman of the AWTSC.

Martin's role in both Hurricane and Totem caused some tension between the two governments, as the Australian Government tried to ensure an Australian presence. In a top-secret letter from the secretary of the Defence Department Frederick Shedden to Allan McKnight written just before Operation Hurricane, Shedden said:

> [t]he Monte Bello trial is possibly a beginning of a series of further trials to be conducted at Woomera, and it is essential to establish the place of the Australian Government Machinery ... The initial step is to clarify the status of the Defence Scientific Adviser ... We are not of course interested in the weapon itself, but only in its effects and the general set-up of the test. It is agreed that Professor Martin should accompany Dr. Penney on a visit to Woomera [just before Hurricane] in view of the probable use of the range there.[18]

In contrast, Titterton's involvement was announced by press release. Additionally, Titterton was allowed access to detailed scientific measurements from the Totem series and made many measurements of his own, whereas Martin, like Alan Butement, chief scientist from the Department of Supply, was an observer only.[19] Titterton was always part of the British camp; Leslie Martin was not.

A small hint at the impatience the Australians were feeling at the British demands comes through in McKnight's closing comment in his letter to Davey, largely devoted to dealing with the British concerns about Martin. Right at the end, the Australians asked for their own assurances of secrecy over the use of Emu Field for atomic tests. 'May I raise one [point of concern] on our side? We would like all communications which mention the possibility of Woomera as an atom weapon test sight [*sic*] to be classified "Top Secret"'.[20] Does this also hint at extremely secret plans for the Woomera rocket tests, apparently completely separate to the AWRE activities at Emu, to be part of the atomic experimentation being conducted in Australia? While the covert activities of Emu and Woomera had some limited crossovers, they were not closely related projects.

Media stories at the time were often wrong or distorted, in large part because misinformation filled the vacuum created by the strict management of information. The Australian media had also signed on, secretly, to self-censor through D-notices. For example, the *Argus* of 3 August 1953 stated that '[a]mong the weapons will be a medium sized atom bomb – Britain's second – which will plunge to the desert floor from an Australian-manned Canberra jet bomber'.[21] Both Totem tests were low-yield and detonated from towers, not dropped from aircraft. This story made the outlandish claim that '[t]he October tests will make Australia one of the world's Big Four in atomic research – with the U.S., Russia and Britain'.[22] This was in no sense true. Australia was not then and never has been part of the club of nuclear armed nations. Even at the time, such a claim must have seemed a laughingly gross over-statement to any British personnel who happened to read it.

The *Age* speculated on the same day the weapon was tested, quoting information from the RAF:

> British scientists are assembling a type of atomic bomb which can be aimed at its target more accurately than any other yet devised. A prototype, it was said, would be tested on a special range being built in the Australian desert. The target area would be presented to the navigator-

bomb aimer on a cathode ray screen in the form of a radar map. The bomb would have no inherent propulsion unit. The bomb aimer's task would be to guide it by remote control.[23]

While this may have been in relation to missiles being developed at Woomera, it was not related to the devices tested at Emu. Blue Danube would never be attached to a British missile. Understandably, because the facts were not made available, Woomera and Emu were conflated in the public sphere and speculation was rife.

These early stories often had incorrect information about what Totem entailed. In July 1953, the Melbourne *Herald* stated that there would be three atomic weapons tests at Emu, involving:

> Detonation of an atom weapon on a 200 ft. high tower.
> Explosion of an atom weapon at ground level.
> Dropping of an atom bomb from a Canberra jet bomber.[24]

This story also asserted that '[t]he new experiments are three months ahead of schedule because of the good progress made in developing the weapons'.[25] This was incorrect, since the only time tests could be held at Emu were in October, and the AWRE had scrambled to fit that timeframe.

Three days earlier the same paper had claimed that six tests would be held at Emu: 'The "big bangs" will be the first of a series of the most intensive atomic weapon experiments ever held. It is expected that there will be at least six explosions between the early autumn (English) and the late spring'.[26] The declassified record does not indicate a program of six weapons tests at Emu was ever actually planned.

Amid speculation one *Sunday Telegraph* report claimed:

> Sir William's unit has been working on two [types] of atomic war-head. The first is designed to develop the full potential of atomic explosive by making maximum use of the plutonium in the war-head. The second is a tactical weapon which can be exploded within eight miles of friendly troops.[27]

At the time of Totem, the populist left-wing Sydney tabloid the *Daily Mirror* started flexing its editorial muscles against the test program, frequently targeting Menzies. The *Mirror* was consistently sceptical about the British test program, largely because of its ideological stance against the Liberal government and its fierce competition with its pro-tests rival the Sydney *Sun*.[28] The *Mirror* was one of the few mainstream news outlets to question Australia's role in the bomb tests, although its arguments were political rather than knowledgeable. A comment piece in October 1953, just before the first blast, presented like a news story under a large photo of Menzies, was typical of its tone:

> Upon his [Menzies'] shoulders rests the responsibility for whatever might follow the forthcoming atomic blast at Woomera. Mr. Menzies has stated that all precautions have been taken, but steadfastly refuses to indicate what the actual precautions are and the type of bomb to be exploded. Scientists in charge of the explosion have repeatedly stated that the explosion will not take place unless there is a favourable wind. In the opinion of this newspaper, this indicates that some risk to the Australian population is involved. If there were no danger, the bomb could be let off

at any time. There is plenty of scope in the Antarctic for such an explosion – not in Australia. Mr. Menzies' responsibility is, indeed, a fearful one.[29]

The *Daily Mirror* was an outlier. More typical was a *Sydney Morning Herald* article on public safety in relation to Operation Totem. This lengthy piece, published six days before Totem I, provided largely unattributed information assuring the public that the tests were safe: 'it appears that the atomic tests on the Woomera range will be even less dangerous than those in Nevada [where the United States was testing its own weapons]'.[30]

The only person quoted in the article, well-known physicist Professor Harry Messel from the University of Sydney, later a critic of the test program, said: 'I do not believe that there will be any danger to the present population', although he emphasised the need – as did the article in general – for extensive monitoring activities (which were not in place).[31] At the end, a small warning note was sounded:

> The atomic bomb to be exploded near Woomera will be the 49th nuclear explosion contrived by mankind during the last eight years. If mankind continues to test atomic weapons at this rate, or at an increased rate, will hazardous levels of radioactivity be built up throughout the world? Many nuclear and radiological scientists, while dismissing immediate hazards as negligible, are not so ready to dismiss the possibility of a long-range hazard.[32]

A follow-up article the next day expanded on the long-term dangers, presenting more detailed information on the hazards of exposure to radioactivity. The article reiterated the need for an effective monitoring effort so the Australian authorities would know the level of radioactivity arising from the British tests.[33]

Indeed, the cumulative effects of the rapid development of nuclear weaponry were starting to cause considerable concern in scientific and political quarters, and subsequently among the general public. The world's atmosphere was indelibly marked by the release of radioactive products created by atmospheric atomic weapons tests, and the bodies of everyone on the planet would potentially bear this signature. Well before the end of the 1950s there were moves to stop the uncontrolled emission of the radioactive products of weapons testing into the global atmosphere, and to measure the radioactivity that had entered biological food chains since 1945.

Between the two Totem tests, Menzies defended Australia's role in the ongoing atomic test series during his regular weekly radio broadcast, *Man to Man*:

> There is tremendous public interest in Atomic Bombs … Unfortunately there are scare stories, wild allegations, and, between you and me, a good deal of nonsense … But we must face the facts. And they are that the threat to the world's peace does not come from the Americans or the British, but from aggressive Communist–Imperialism. In this dreadful state of affairs, superiority in atomic weapons is vital. To that superiority Australia must contribute as best she can.[34]

The guidelines prepared for the press in the lead-up to Totem I told journalists they would view the blast from a ridge near the base camp and be able to see the top of the 30-metre tower. They would be given dark welding glasses on-site for eye-protection, so they could watch the initial fireball, although these would not be much use once the fireball lost its brilliance. They were advised: 'Observers may wish to look at the sun through the [welding] glasses a few seconds before explosion to get a comparison of the brightness of the sun and the brightness of the fireball'.[35] The other option was to do the same as on-site personnel:

> As the count five-four etc., starts, to turn their backs on the explosion, and wait until they have heard over the Loudspeaker 'ONE' which follows, repeat follows, 'ZERO' (This means that they must hear FIVE-FOUR-THREE-TWO-ONE-ZERO-ONE BEFORE THEY TURN ROUND).[36]

Journalists travelled to Emu in an aircraft with its windows covered in canvas to obscure the exact location of the test site, no doubt adding to the air of mystery and adventure.[37] They were previously quartered for 12 days at Salisbury, due to the week's delay for Totem I.[38] Ronald Monson reported:

> On reaching the site we were taken by truck to our observation point, where, to the astonishment of the English correspondents, at least, a breakfast of curry and rice and Scotch whisky awaited us in a tent. I noted that the English correspondents hardly touched their curry and rice.[39]

The day after the first shot, the *Sydney Morning Herald* had a lead story on its front page with sidebar stories and a photo series showing the mushroom cloud rising from the desert plain. The headline, 'Atom Explosion Success', set the tone. The article was written by 'A "Herald" special reporter, who watched the atomic explosion on the Woomera Range from

15 miles away', a unique by-line. The story was partly reported in the first person, which was then unusual in a lead news story: 'I chose to turn my back at the initial explosion, not wear glasses, and look around on the count of one [second] after zero'.[40] This special reporter provided some vivid imagery of the test:

> At zero hour, the flash lit up the sky, despite the bright, sunny Central Australian morning. In the first two seconds, a huge ball of fire rose about 750 feet, oxides of nitrogen forming and remaining in the air. Colours, mainly brown and red, flamed. In these sweeping two seconds, as the flames shot upwards, several of the observers noted the face of an Australian aboriginal formed by the soaring flames.[41]

The entire paper was dominated by bomb test news that day. A news story from a 'special correspondent in London' provided a drawing of the Totem bomb just exploded. This brief item had limited information and made a pertinent point: 'It is emphasised that this is only the principle of the weapon which is already internationally known. Atom spies conveyed it to the Russians. Refinements embodied in the British bomb, which greatly increase its efficiency, are still secret'.[42]

The issue also featured an interview with supply minister Howard Beale, who was quoted in the story saying: 'One of the most pleasing aspects of this project is the proof it provided of the effectiveness of Australia's security arrangements'.[43] This was an important line for the minister, as Australia was still working behind the scenes to assure the sceptical Americans that it was sorting through previous security problems that had led to the establishment of the Australian Security Intelligence Organisation a few years earlier. The security breaches were strongly associated with the previous Chifley Labor government

– and indeed with Australian science – and the conservative Menzies sought to rectify them. Beale emphasised: 'There has not been a single security leakage in Australia during the months of large-scale preparations, and many Australians worked on the project without knowing the purpose to which the range would be put'.[44]

The article addressed at some length the Australian contribution to the short-lived test site at Emu Field, touching on the hardships the workforce faced there and concluding:

> Mr. Beale said all Australians had cause to be proud of their country's contributions to 'this portentous event. Although only a small nation, we have given eloquent proof of our ability and willingness to take a major part in the defence of the freedom of mankind', he said. 'But we also pay especial tribute to our great United Kingdom partner who, in spite of her enormous other commitments, has so magnificently contributed her best scientific brains and technical skills, as well as much equipment and many millions of pounds, in order to bring this project to such a triumphant conclusion.'[45]

In an article rounding out its coverage of Operation Totem following the Totem I explosion at the end of October, Menzies said that the second test 'confirmed theoretical calculations', while Beale, again interviewed, noted: 'It has been a very big job, and its successful conclusion reflects great credit on everyone associated with the tests'.[46] Most notably, this edition of the *Herald* included a brief item on minor tests to be conducted at Emu Field, the first five Kittens experiments, one of the few media mentions of these under-the-radar tests. The story came from Australian Associated Press and was based on a media announcement by the UK supply minister Duncan Sandys, who was quoted saying, 'During the last few weeks our scientists have carried out various other trials, involving a number of minor explosions'.[47] No other information was given, other than to say that the test program was now complete.

British journalist James Cameron, writing in the *Age* after the first blast, sounded an ominous note: 'The familiar mushroom column climbed unsteadily for 15 000 feet, leaned and dropped and the world stumbled one more step towards the twilight'.[48] Cameron's story included arresting descriptions of the event, including the moments leading to the detonation:

> The count down through the loud speaker emptied the seconds away in the thin precise and very English tones of some disembodied scientist. He said it in a detached metallic way, '– four – three – two – one – zero –' and then he had no need to say more.[49]

The column of the Totem I cloud was a dirty brown, 'somehow dripping a little curtain of vaporous grey at one side', in contrast to a US test he had witnessed at Bikini Atoll, which was like 'sculptured raspberry ice cream'.[50] Was this revealing of the black mist that was to follow this blast?

After his rousing editorialising about the Totem bomb, Cameron added a note at the end of his story, titled 'After Thoughts on X200', perhaps to calm anyone concerned about what he was describing:

> There will almost certainly be a measure of disappointment over our X200. It did not after all blow a hole through Australia; the first guess that it might be heard 200 miles

away is almost certainly exaggerated; no one was even jarred. As far as I know nothing has dropped off any of us yet. Just as the city of Honolulu protested in fear at the Bikini test, and then groused that there wasn't a tidal wave, so Adelaide and Melbourne may well feel cheated that nobody had to duck.[51]

Somehow it seems unlikely that anyone in Australia was disappointed they were not harmed by Totem I.

The secret Defence, Press and Broadcasting Committee issued a D-notice for the Totem tests at Emu Field in October 1953. It forbade reporting on a range of issues, including '[n]uclear efficiency and measurements relating to weapon efficiency' and listed a number of side issues that would be provided to reporters as background material, including:

(a) Initial survey of the area.
(b) Survey by Sir William Penney.
(c) Work of construction personnel.
(d) Assistance given by L.R.W.E.
(e) Air-lift operations by Yorks and Bristols ...
(f) Boring operations for water and study of geology.
(g) Work of Australian scientists in checking margin of safety.
(h) Transport of aircraft and war stores to site.
(g) Co-operation of pastoral lessees.[52]

On 26 June 1953, just before this D-notice was issued, Arthur Fadden, the acting Australian prime minister, sent a letter to the press representatives on the Defence, Press and Broadcasting Committee. At that point the media was barred from Totem, with Fadden's full endorsement, to eliminate media pressure on the test authorities:

It is desirable that the man in charge of the operation should have a considerable margin of time to play with as to when the test should take place. The presence of the press has a tendency to lead to attempts to meet a scheduled date and this could cause a reduction in the value to be derived from the test.[53]

Fadden was echoing the general views of both governments that the media were a distraction and a pressure. In a letter to Neil Pritchard in January 1953, Ben Cockram discussed the ways in which the media were not only kept in the dark, but were also mocked:

Bowen instanced, as an example of the way in which the Press had been treated at Woomera in the past, that they had never been allowed to see anything except the amenities of the Camp, finishing always with a visit to the sewage works as the 'piece de resistance'. South African journalists, when they had been taken to Woomera and given this routine had been sharply critical of the waste of their time involved and had, in fact, refused to get out of the cars and inspect the works at all.[54]

Hearty laughs were had by all.

Menzies constantly and possibly defensively asserted the safety of the Emu tests, not that too many people were indicating public concern. He was quoted in the *Age* in September 1953: 'the tremendous importance of the Woomera experiments was that they were calculated to save lives and not imperil them. Scare stories about this vital and important work only made it more difficult'.[55] Menzies also made clear that his government 'would have nothing to say at any stage about the "experiment or experiments whichever it might be'.[56]

The Australian media was often interested in the apparent secondary cold war between Britain and the United States when it came to nuclear weapons development, and speculated about any potential US involvement at Emu Field. The *Sunday Sun & Guardian* revealed that American representatives were possibly going to attend the tests. 'If they do, it will begin a new era of rigidly controlled co-operation between Britain and America in nuclear science.'[57] The article claimed that the question of US representatives attending the tests was the subject of considerable diplomatic activity. 'High Federal authorities say Americans were surprised at the progress demonstrated by Britain at the Monte Bello test, off Western Australia, last year. Americans are keen to know more details of British progress. Federal officials believe the Woomera tests may prompt America to exchange information.'[58]

The British authorities loved to blame Australia for security and media leaks, and did so often. This tendency was discussed in a top-secret letter from Jack Stevens to Allan McKnight in May 1953. The *Daily Mail* in the United Kingdom had published some information about the forthcoming Emu tests, causing much consternation. The article was written by a journalist called J Stubbs Walker (who appears most notable for his stories on UFOs). TOTEX head Thomas Elmhirst had written to Stevens, and rather passive-aggressively put the blame on Australia: 'Our Information Branch guesses that Stubbs Walker got his idea from Australia but in fact we have no knowledge of where his background information came from and were as surprised as you must have been to read the story in the morning newspaper'.[59] Stevens picked up on the jab in his letter to McKnight, saying:

Note the gentle reference to 'Stubbs Walker' having got his idea from Australia!! It's an old British custom, I'm afraid. If Elmhirst were here, I would take it up with him but I'm doubtful if it's worth fighting about. Unless, therefore, you have any other news I propose to tell our Panel here that we will ignore the barbed shaft.[60]

The secretary of the Prime Minister's Department, Allen Brown, wrote to Prime Minister Menzies on 28 September 1953 to discuss the finer points of how the tests were to be discussed in public. He dropped Howard Beale in it by noting that Beale had been quoted in the *Sydney Morning Herald* that day talking about 'Australian observers' at the test. As Brown told Menzies:

> So far we have been careful to avoid any reference to the term 'observers' and have referred to scientists with a job to do. Professor Titterton definitely comes within that category and I believe Professor Martin and Mr. Butement are given specific duties so that they rate as 'working scientists'. If the press mentions this to you, you may care to shift the emphasis back on to 'working scientists'.[61]

This appears to be part of the overall attempt by the Australian Government to give the impression that Australian scientists were at the heart of the tests.

In all, 17 representatives of the UK and Australian press witnessed Totem I from a site to the west of the camp and about 25 kilometres from the tower holding the device, some distance from the lookout where Penney and other insiders were standing.[62] These rather unwelcome visitors were quickly ushered away after the detonation, without being allowed to visit the camp. Penney said: 'The Press observers have been given permission to take still camera photographs and without censorship to describe the explosion as they saw it'.[63] They were also provided with heavily vetted information about what they were seeing.

The contemporary media were not oblivious to the connection between the events at Emu Field and the financial implications of Australia's bountiful uranium reserves. An editorial in the *Sydney Morning Herald* a week before the first Totem shot explicitly connected them.[64] Titled 'Atomic Bombs, Uranium and Australia's Future', the editorial began:

> This is a momentous week for Australia. To-day or tomorrow [in fact, seven days later] the first atomic bomb will be exploded near Woomera with half the world's press looking on – at a discreet distance – and with all the apparatus of publicity. To-day or tomorrow also, with no publicity at all, Lord Cherwell [Lindemann] will be meeting Federal Members at Canberra to negotiate a long-term agreement for the sale of uranium to Great Britain. These two events are of course closely related.[65]

The editorial went on to make a case for Australia to be less of a source of resources for the United Kingdom and more of an atomic power in its own right. It concluded:

> Should it not be part of the agreement ... that Britain should help us to build a uranium refinery and an atomic pile in Australia? ... Otherwise, we may find ourselves committed for years to come to shovelling uranium ore into ships for England while we fall further and further behind the rest of the world in technical skill. Australia is not

like Saudi Arabia or Persia, who sell their oil because they cannot use it themselves. We may have enough uranium for export as well as for our own use, but it is time to think of the future.[66]

The select few journalists who witnessed Totem I were loyal to the Mother Country. Ronald Monson from the *Daily Telegraph* was among them. He wrote on 24 October:

> None of us wanted to give anything away to a potential enemy, but, of course, there was the danger that we might have divulged some apparently innocuous information that the boys on the other side of the Iron Curtain would have found useful. We needn't have worried. We never found out anything we weren't supposed to know – even by accident.[67]

Triumph, indeed rapture, often featured in news about the early British tests in Australia, although later in the 1950s the tone became less enthusiastic. Most media who covered Totem were laudatory of Britain's testing activities in Australia. Sydney's *Sunday Herald* ran an upbeat editorial that expressed confidence that safety measures were in place:

> Whether any risks at all are involved in the Woomera atomic explosions is a point on which even scientific opinion may differ, though the more romantic and alarmist versions of the possible A-bomb tests can be dismissed. After all, a number of bombs have been exploded in the United States far closer to centres of population than is Woomera.[68]

As ever, these media glimpses of the bomb tests were through the lens of Westerners unfamiliar with Anangu presence in the lands. Only much later did anyone in the West think to question with any depth or persistence the effects of these uncontrolled experiments on human beings.

Chapter 9

AUSTRALIA'S ATOMIC BARGAINING CHIP

> While it would be comforting to believe that all barriers between the countries concerned are to be demolished and that atomic information and assistance will flow freely, the history of the recent years raises some doubt as to how real the co-operation will be, at least for some years to come.
>
> Howard Beale to Percy Spender, Australian ambassador to the United States.

One of the intriguing political undercurrents at the time of the Totem tests concerned uranium – the material that powers atomic weaponry and energy. The three-way politics between the United Kingdom, the United States and Australia was a diplomatic tussle between the two major powers to woo the minor power into supplying their needs, not just for bombs but for electricity. Plutonium is made in a reactor (it is not found in nature) by bombarding natural uranium with neutrons. Therefore, while uranium itself was generally not used in the atomic weapons tested in Australia, it was needed to create the fuel that powered them. While uranium, a heavy element, is not plentiful, some countries have significant reserves, and Australia is one of these. The country found itself in an interesting position at a time when the two major Western atomic powers were in the market. The other known reserves in the 'Free World' were in the Belgian Congo, Canada, South Africa and the United States (the Soviet Union found its own sources within its territories, principally Kazakhstan). Most of their uranium had been committed to the weapons programs of the United States and United Kingdom.[1] However, a lot more uranium would be needed to develop atomic energy. In the 1950s, both military and commercial imperatives were in play, and they were not always compatible.

The Australian Government first learned of the military uses of uranium in 1941 when the Australian physicist Mark Oliphant briefed it on one of the biggest secrets behind closed Allied doors.[2] Oliphant was centrally involved in the emerging science of atomic weaponry in the United Kingdom as part of his storied physics career and was later part of the Manhattan Project. He disclosed plans to build an atomic bomb to the Australian ambassador to the United States, Richard Casey, while in Washington in 1941.[3] At that time, he told Casey about the work of Britain's Maud (Military Application of Uranium Detonation) Committee, work that was later transferred to the Manhattan Project. During his brief trip to Australia in 1942, when he also informed the government about the new technology of radar, Oliphant likely discussed the doomsday weapon being created by the Manhattan Project with David Rivett, head of the Council for Scientific and Industrial Research, forerunner to the Commonwealth Scientific and Industrial Research Organisation, an organisation to which he had been seconded.[4] Oliphant had been in regular contact with Rivett and had discussed Australia's uranium resources with him, in light of the growing military need for uranium ores.

In the early 1950s, Australia had two known uranium deposits, at Rum Jungle in the Northern Territory and Radium Hill in South Australia, and there were favourable signs there was more to find.[5] Rum Jungle began production in September 1954. Radium Hill had been active since 1898, when radium (rather than uranium) was extracted for a niche market in some medical and 'beauty' treatments.[6] Most of the output of both of these mines was earmarked for the UK defence program for seven to 10 years after an agreement on Rum Jungle was signed in January 1953, via the Combined Development Agency (CDA).[7] This was an Anglo–US cartel, originally called the Combined Development Trust, set up in 1944 as part of the wartime Quebec Agreement in a (vain) attempt to monopolise sufficient quantities of the world's uranium resources to meet growing military needs. For civilian energy purposes, both Australian mines were expected to have some surplus uranium oxide (yellowcake) after the extraction of the military stocks. In addition, a huge uranium deposit at Mary Kathleen in Queensland was exploited from 1954 until the 1980s.[8]

Australia, then, was fortuitously in possession of an extremely valuable commodity as the atomic age burgeoned. This became especially salient when the United States considered making small changes to its controversial McMahon Act so it could share some atomic information with Australia, in exchange for a contract for its uranium.[9] The move infuriated the British, because one of the stumbling blocks as it worked to overcome the hated strictures of the McMahon Act was American wariness of its trade agreements with 'the Dominions', including Australia.[10] American concern that these agreements might oblige Britain to share atomic secrets undoubtedly influenced British behaviour towards Australia in offering to share knowledge as an incentive for atomic testing on Australian soil, without actually doing so. Now America wanted to supply or withhold that largesse on its own terms.[11] If America snatched away Australia's resources, Britain would be forced to look at other possible uranium supplies, and turn its attention to various countries in Africa.

Reliable uranium supplies were critical in a post-Hiroshima world. British physicist Sir James Chadwick was among the UK insiders who pushed to have this understood as key to decision-making around the future atomic tests: 'The chief fact we have to bear in mind is that we have no stocks [of uranium] in England. I am weary of pointing this out and of urging that we should take steps to remedy it'.[12] The United Kingdom had undertaken an extensive survey of its own territory trying to find the newly valuable element. It discovered some deposits in Cornwall, but not enough to be viable, making supplies from other countries essential. As Margaret Gowing put it, in her inimitable style:

> In the formulation of policy towards Europe and the Commonwealth, uranium was the criterion which divided the sheep from the goats; irrespective of any wider political considerations, countries were treated more or less favourably according to whether or not nature had endowed them with uranium-bearing rocks.[13]

UK minister of supply Duncan Sandys told parliament that the 'decision to rely on uranium deposits overseas is ... not a question of policy, but of necessity'.[14]

Australia was about to find out quite a bit more about this necessity when the British sought to secure quantities of Australian uranium as part of overall negotiations to test weapons there. In 1950 the Australian Defence Committee reiterated that searching for uranium deposits should be considered a national priority, stressing 'the *military* importance of instituting a thorough search for uranium in Australia and its territories'.[15] The department had been authorised since 1946 to offer a reward of up to £25 000 to anyone finding uranium in Australia, to encourage prospectors to get their Geiger counters out.[16] (The reward was paid out to prospector Jack White, who discovered the deposit at Rum Jungle in 1949.[17])

In 1952, Frederick Lindemann briefed Churchill on the options:

> South Africa will soon be one of the biggest producers of uranium ore and Australia will be another. Australia would undoubtedly like to work with us, but we have hitherto been hanging back so as not to jeopardise our chance of cooperation with the Americans who have seemed to be jealous of our working with anyone else.[18]

Lindemann, deeply suspicious of American motives, was prominent among the British officials who encouraged Churchill to counteract US moves to monopolise nuclear technology. As the United States voraciously pursued sources of uranium around the world, Lindemann believed the United Kingdom had a built-in advantage through its associated Commonwealth and should exert its influence to provide a counterweight to US might. Britain even attempted to claim that Commonwealth uranium resources were actually *domestic* resources. Under the CDA military uranium agreement, the CDA could commandeer all uranium resources discovered in the so-called 'Free World' for military purposes. However, the territory of the United States and British Commonwealth was exempt from this provision. So if the United Kingdom claimed Commonwealth resources as 'domestic', America and the CDA would not have access to them. As it turned out, Britain was forced to abandon this position 'simply because it lacked the financial resources to develop these sources'.[19] Since Britain could not freely exploit Commonwealth resources, it had to negotiate with the Commonwealth countries. The United States was not going to help. In the note to Churchill, Lindemann was ever sceptical about getting along with the Americans at all:

> I do not think it would be worth while to make any serious sacrifices for the sake of getting only limited cooperation with the United States which, I fear, is all we are likely to obtain. Unless we can get really full cooperation I think we should seriously try to build up a great Commonwealth enterprise to develop the industrial and military uses of uranium. In this the Commonwealth would find the material and we (at any rate in the beginning) would make the main technological contribution. Whether the Commonwealth countries would agree to this, or whether they would succumb to the attraction of American dollars is uncertain. But I am inclined to think it is our best card.[20]

He was also sceptical of Britain's value to the Americans:

> Quite frankly we have very little to offer the Americans. We believe that some of our processes are rather more economical than theirs, but they care little for this. What they really would like would be some of our best men whom they will no doubt try to lure away.[21]

Lindemann pressed hard on the idea of loyalty among the Commonwealth nations – if they had uranium – conceding that while Britain could obtain some good technical information from the United States, it was 'not essential'.[22] America had already exploded some 30 atomic weapons and was close to exploding a thermonuclear device, a weapon that he said was at that point 'quite beyond our means'.[23] Since the Americans knew that Britain was not in the H-bomb race at that time, there was no prospect of any co-operation with them in that direction.

Lindemann concluded his note to Churchill saying he 'would therefore, today rather get the uranium than the information', and was deeply concerned that the Americans would get to Australian uranium before the United Kingdom did.[24] In a message to Churchill on 4 May 1953, he reiterated the source of his concern:

> The danger is that the Australians may give the Americans a monopoly of their ores in return for promises of help which may well turn out illusory. It is in order to discourage this that I put in the draft telegram to Menzies, the rather emphatic phrase:- 'It does not seem likely that anyone will ever be able to get any really useful collaboration with them'.[25]

Lindemann wanted to plant the idea in Australian minds that American offers of technical assistance would come to nothing, something that, as it turned out, applied equally (if not more) to British ones:

> What we would like of course would be an agreement with the Australians by which they would sell ore to us in return for technical help. But we hope to achieve this

without making it an absolute condition at this stage; and it is thought that, if Menzies is prepared to work with us in principle, the best thing would be to discuss details with his officials who are coming here shortly in any case.[26]

Australia was not entirely unconscious of the valuable bargaining chip it held in its diplomatic game with the United Kingdom if it wanted to have a role in atomic research and development. In 1952, Britain started to become worried about counting on Australia's continuing compliance.

> Ironically, in view of Prime Minister Robert Menzies' extraordinarily accommodating attitude over British nuclear weapons testing in Australia (soon to be seen in the detonation in October of Britain's first device, off the Monte Bello Islands), his government's position on uranium development proved to be unpalatably independent for London's tastes.[27]

In fact, when Menzies travelled to the United States in 1952, uranium was one of the issues on his agenda. He met with the US Atomic Energy Commission concerning the development of Rum Jungle, then about to start production.[28] While Australian officials at the time repeated the mantra that the country's relationship with the United Kingdom came first, the approaches from the United States were attractive. Devoid of the complications of Australia's post-colonial relationship with Britain, the Australians enjoyed the openness of the Americans, in contrast to the closed, cautious and condescending Brits.[29] Britain was being particularly cautious with Australia as it worked to re-open atomic co-operation with the Americans.

In May 1953, the Australian minister for supply, Howard

Beale, wrote to his prime minister, detailing Australia's atomic role and ambitions. The letter was sent in light of the creation of the Australian Atomic Energy Commission the previous year and the growing realisation of the extent of Australian uranium resources. Beale pointed out that the South Australian premier, the Liberal Tom Playford, had been so enthusiastic about his state's uranium resources that he had made direct approaches to the United Kingdom, Canada and the United States to establish export agreements. These representations

> have given rise to some doubt in the minds of the atomic energy authorities in those countries as to what organization in Australia they should deal with in the field of atomic energy research. I have always held the view ... that it would be infinitely better for Australia to speak with one voice in this connexion and as a result of recent developments here ... it seems that this desirable position might soon be reached.[30]

Playford argued vigorously for the development of an 'atomic pile' (reactor) in South Australia to supply the state with energy. South Australia did not have sufficient coal resources, or enough water for hydro-electricity, and Playford wanted his state to be energy self-sufficient by 1960. However, Beale didn't want any state controlling this new technology or its raw materials and sought a coherent national approach led by the federal government. Beale advised Menzies that while there was 'much to be said for developing a joint project with South Australia ... the Commonwealth [would have] the paramount authority'.[31]

In setting the terms of forthcoming negotiations with the UK government later in the year, Beale articulated Australia's bargaining position:

In any agreed programme, Australia should be admitted to equal partnership with Britain, and all information, reports, etc., should be made available to us. Britain will probably wish to exclude purely defence information, i.e., how to make an atomic weapon, and this would seem to be reasonable, at least until we have stocks of our own plutonium.[32]

Beale thought the United Kingdom should be aware that Australia wanted to keep open options for selling uranium to the United States.

The United Kingdom could not advance its military and civilian atomic ambitions without uranium. The two main British ministerial decision-makers on these matters were Frederick Lindemann and Duncan Sandys, a pairing that guaranteed conflict: 'Whether out of insensitivity or mischief, Churchill had upon returning to power installed his son-in-law Duncan Sandys as his Minister of Supply, long one of Lindemann's arch-enemies in Whitehall'.[33] Their antipathy dated back to World War II, when they were on either side of the 'Scientific Controversy' concerning intelligence reports about the development of German rockets. Sandys believed the rockets posed a threat to the United Kingdom, while Lindemann argued they were just an elaborate cover story.[34] They clashed again when Lindemann pushed for the creation of the UK Atomic Energy Authority, removing responsibility for atomic matters from the Department of Supply.

Both men headed to Australia in 1953, close to the start of Operation Totem, for high-level discussions with the Australian Government. Sandys preceded Lindemann, arriving on 30 August 1953. On his arrival, he told a media conference at Sydney's Mascot airport that he would not be staying for the

Emu test, and did not know exactly when it was going to be held:

> 'It's impossible to tell the exact date now', he said. 'It depends on meteorological and other conditions. Every precaution must be taken that winds of the right strength are moving in the right directions. You can best guess the date the tests start by wetting your finger and holding it in the air.'[35]

Nevertheless, his mind was on the first Totem test as he entered into discussions with Menzies. As minister for supply, Sandys was in charge of equipping and supplying the armed forces. He discussed the ongoing guided rocket research at Woomera and the atomic test program – both the forthcoming Totem test and early plans to establish a permanent atomic test site at Maralinga. He also raised the prospect of increasing uranium exports from Australia to Britain, trying to soften up the Australian prime minister ahead of harder negotiations with Lindemann.

Sandys subsequently visited the Emu Field site and soon after sent a note of thanks to William Penney: 'I was greatly impressed by what I saw of your vital work and would like to congratulate you and your team on the rapid progress you are making in preparing for next month's atomic tests.'[36] He also visited Radium Hill and met with Playford, which the *Sunday Sun* suggested had gone well:

> Reports from Adelaide say that South Australia may get more British capital following the visit to Radium Hill yesterday of British Supply Minister Duncan Sandys. He will leave Australia with assurances by South Australian

Premier Playford that South Australian uranium production could be increased rapidly if the need arose. After visiting Radium Hill, Mr. Sandys said the British Government set great store on the work carried out there.[37]

After three days touring Woomera and the LRWE facility at Salisbury, Sandys commented for a Melbourne *Herald* story:

> 'As British Minister of Supply, I am in continuous touch with the progress of guided – weapon development in the U.K., which is making remarkable strides', said Mr Sandys. 'But until I came here I frankly did not realise fully the great work which was going on in Australia in this field. Not until I personally saw Woomera this week did I appreciate the true state of this far-sighted project, and the full extent of Australia's contribution to the joint defence effort. Within a few years the Australian authorities have created in an almost uninhabited area extensive installations. These, coupled with the knowledge in inventiveness and enthusiasm of both Australian and British technical staff, as well as the airmen and administrative staff, together make a splendid team to make Woomera the finest weapons testing ground in the world.'[38]

Buttering up the Australians wouldn't do any harm for the bigger game of uranium negotiations. A few months earlier, when Menzies had visited Britain for discussions on uranium, Lindemann took the lead and attempted to reel Menzies in so that he would not be beguiled by America. Initially, Lindemann appeared to believe that he had Australian uranium in the bag. He briefed Churchill in early June:

> Since the arrival of the Australian party there have been discussions between officials and I understand that Menzies is quite prepared to give us the assurance about uranium supplies that we are anxious to obtain. Furthermore, while he would no doubt welcome a brief word of thanks from yourself, he is quite willing in view of the very many calls upon your time, to discuss the next step direct with me.[39]

Quickly, though, he hit a snag. Uncharacteristically, Menzies did not roll over immediately. The Australian prime minister had become frustrated because he perceived Lindemann was over-optimistic about how much uranium had still to be found.[40] Menzies wanted to disabuse him of the notion that it was a simple matter of just going out into central Australia and digging it up. He suggested the postmaster-general travel to Australia to better understand the issues. Lindemann did not want to go to Australia, but progress in the negotiations with Menzies in London was too sluggish for his liking:

> Our discussions with the Australians about cooperation in developing the industrial possibilities of atomic energy have gone well but more slowly than we hoped. It is now clear that Menzies will not give us a definite answer until he has returned [to Australia from Britain] and put the report of the Australian exports before his Cabinet. He seems to expect that I should go out to Canberra for a second round of discussions, probably in September, when a final agreement would be reached.[41]

Churchill, by hand, circled the word 'Canberra' on this note and wrote 'Oh Lord!' beneath it. The very thought of having to visit the Australian capital was distasteful. This undertone of contempt is occasionally palpable in the documents. But Lindemann's aim was 'to try to barter Britain's entire civil atomic knowledge for Australian supplies', so Canberra it would have to be.[42]

Lindemann arrived in Australia in October, at almost exactly the same time as the Totem devices landed.[43] While in Australia, he made a quick trip to Emu Field, accompanied by Alan Butement.[44] He was in poor health, a state of affairs not helped by the brick wall he encountered.[45] He had wanted to monopolise the uranium resources in the Northern Territory and South Australia by creating a miniature civilian version of the CDA, but Australia refused. The Australian Cabinet took the view that it had discharged any obligation it might have to its former colonial rulers by allowing the British to test their atomic weapons on Australian soil. These obligations did not extend to supplying uranium.[46] The talks collapsed, just as Operation Totem was about to begin. After three days of discussion, Lindemann left without an agreement. According to scholar Alice Cawte, '[h]e made one last plea that Australia at least inform the United Kingdom when it intended to sell more uranium to the USA. Cabinet flatly refused'.[47] Australia rejected Lindemann's proposals, saying the United Kingdom would be Australia's 'preferred customer for any exportable surplus of uranium products remaining after satisfying our own requirements and making such contributions to the defence of the free world as was considered expedient'.[48] Australia stood firm on this issue, perhaps because of the significant economic earnings on offer.

Media pressure was building on Menzies to do a favourable deal for Australia, while nodding to the ties that bound. An editorial in the *Argus* spelled this out:

> we would rather share the venture with Britain than with any other great power in the world. But if it is to be a true partnership, Australia cannot place herself in the position of the colonial supplier of raw materials. Partnership involves sharing, and before we decide how to share our uranium with Britain, we must know how much we have and how much each country may need … No newspaper in Australia has been stronger in support of the Anglo Australian partnership than *The Argus*, and we believe that it is in the interests of this partnership that Australia should develop into a great industrial power in its own right. In the meantime, let us give as much direct help to Britain – in the form of uranium – as we can reasonably afford. A fair bargain makes firm friends.[49]

The harder bargains around uranium stand in contrast to the pliant agreements Australia made on atomic weapons testing.

A blistering, anti-communist rant of an editorial in the *Bulletin*, not an unusual occurrence, proclaimed: 'In view of the fact that atomic defence is dependent on the know-how of the British and the Americans, both as to bombs, weapons and machines to carry them, it would not be too dear a price to pay as our share if we *gave* our allies our uranium.'[50] Loyalty was not about to win out over trade economics and domestic strategic considerations, however.

In rough notes prepared immediately after the 6 October 1953 meeting of the Cabinet Committee on Uranium, which Lindemann attended in Canberra, a senior official, most likely Allen Brown, set out the reasons for the Australian hard line:

> We cannot accept U.K. request for two-thirds of our production because:-

(1) We don't know much we have, if any.
(2) We must satisfy our own requirements before selling any abroad.
(3) We don't know what our own requirement is (Our knowledge could be delayed if U.K. fails to give us technical information).[51]

Brown said in a confidential note to Menzies after Lindemann had attended the Cabinet Committee on Uranium meeting:

> do not let us be rushed in this matter. This very big scheme of turning over the whole of the power production in the United Kingdom from coal to uranium has never been to the British Government for consideration. It has been sketched in outline only to the British Treasury. The British Treasury has given it only the faintest of blessings and has furnished Lord Cherwell [Lindemann] with just sufficient bait to enable him to go on a respectable fishing expedition.[52]

The *Daily Mirror* reported that Lindemann departed Australia 'in a huff, after failing to pull off what the Americans would call "a very fast deal"'.[53] Back in London, the Cabinet office was processing the unexpected outcome of Lindemann's abortive trip to Australia. A top-secret message sent to Lindemann's office said:

> It seems clear that until Lord Cherwell has returned to England and the Committee has had a first-hand account of his negotiations, it is not practicable to take any decision on the following points: (a) Whether we should try again with the Australians and what we should say to them about their

request for technical help; (b) Whether we should attempt an approach to the South Africans.[54]

The implications for planning were potentially immense, including the issue of sharing technical knowledge:

> We have not yet decided that we want to be free to give classified information to the Australians, and therefore there is no point in upsetting the Americans unnecessarily ... In any case in our original plan we did not propose to open up with the Americans unless we had first assured our base in Australia, which we have not been able to do.[55]

The United Kingdom would have to make some concessions to the Australians, or see that yellowcake disappear over the horizon to America.

In a letter to Churchill in November 1953, Menzies made it clear that Australia was not quite ready to hand over its uranium, as the means to exploit the resource were still gearing up:

> As we have not yet reached the point of production of uranium in Commonwealth Territory, we felt that there were great difficulties about making a specific contract for the future. It will take us a few years to perform our obligation to the Combined Development Agency for military purposes. But we were able to assure him [Lindemann] that when we reached the point of uranium production for civil purposes we will, of course, regard the United Kingdom as a preferred customer in respect of any exportable surplus.[56]

Menzies recounted Lindemann's view that the United States was less concerned with atomic energy, since it had 'abundant alternative sources of power'.[57] He also showed where his loyalties lay: 'The basic fact is that we stand or fall together, and that Great Britain will no more need to worry about Australian co-operation in the future than she has in the past'.[58] He warmed to the theme, adding, '[t]he longer I engage in public affairs the more convinced I am that we must at all times nourish our ancient structural unity which remains the best thing in the free world'.[59]

In March 1954, Menzies wrote a secret, personal letter to the marquis of Salisbury (whom he addressed by his nickname, Bobbety). Menzies was close to Salisbury, who was lord president of the council and Leader of the House of Lords. The Australian wanted to clarify the information-for-uranium deal Lindemann had tried to negotiate with the Australians:

> Arising out of the Prof's visit last October we have now exchanged notes with you through your High Commissioner here, which constitute an arrangement whereby you will make technical information available to us and will train our personnel in your establishments. (We, on our part, will do some experimental work in selected and small sectors in Australia; but this is a relatively minor consideration in view of the great value to us of what you have offered to do.) However, the arrangement is, in effect, in suspense until you obtain the approval of the United States Government for releasing the information to us and we have left this aspect on the basis that the United Kingdom Ambassador in Washington has complete discretion to make his approach to the United States authorities at the most opportune time.[60]

Allen Brown, who drafted the letter to Salisbury, sent it to Menzies with a note that read in part:

> If you are to announce publicly what the British are prepared to do for us, I am sure that British would want to announce our somewhat vague promise to consider, sympathetically, ensuring them of long term supplies of uranium fuel. And that could set off another chain of local criticism of bartering away the country's heritage.[61]

Australia was constantly drawn back to the United Kingdom, but it coveted US dollars:

> While we should treat the United Kingdom as a preferred customer we must be careful of our relationship with the United States ... the United Kingdom would probably take anything we could offer as surplus to our requirements but I think the door should be kept open at all times to the United States.[62]

These issues came to the fore in December 1953 when President Dwight Eisenhower proposed a new United Nations agency to provide a bank of uranium and fissile materials that could be allocated for peaceful purposes, mostly civilian power.[63] Australia danced around both the major powers as it determined its best advantage. Ultimately, Australia would sell uranium for civilian purposes to both, and subsequently to other countries, but it became a difficult political issue domestically for successive governments. Uranium did bring a modest income for a brief period but it was never the anticipated economic bonanza.

While in Australia, Duncan Sandys was able to view first-hand the advantages of testing on Australian soil, particularly if Australia made good on its ambition to become an atomic nation. As Neil Pritchard at the Commonwealth and Foreign Office wrote to Ben Cockram at the UK High Commission in Canberra, Britain wanted to ensure that Australia did business only with the United Kingdom:

> looking some way ahead, we certainly wanted the Australians to look to us, and not to the Americans, for information about atomic weapons as well as about industrial uses. Moreover, if the Australians were to build a plant for production of atomic energy for industrial use, plutonium would be produced as a by-product and would be available for military use. This might prove to be a valuable source of supply for United Kingdom atomic weapons.[64]

The agreement that governed the British nuclear tests in Australia, like any such agreement, was never truly undertaken for noble reasons. Despite any associated propaganda, self-interest always dominates. Pritchard foresaw an atomic-powered Australia that adopted British technology, creating a market for British ingenuity:

> All this was a further argument in favour of going as far as possible with the Australians over industrial use at the present time in order to ensure that any plant they might build should be a United Kingdom type and based on United Kingdom research. This aspect will be mentioned when we make a submission to Ministers recommending that we should go ahead with the Australians over the exchange of technical information for industrial use even though we have no firm commitment about uranium supplies from them.[65]

The British-designed PIPPA reactors that were going to produce both reactor- and weapons-grade plutonium at Calder Hall could be sold to countries such as South Africa, India and Australia.[66]

Pritchard concluded his letter with a plea to Cockram: 'we should be grateful if you would keep your ears open and let us know if ever the Australians should begin to wish to take a real interest in the weapons side'.[67] The Australians showed considerable interest in the weapons side during the 1950s.[68] The agreement struck between the United Kingdom and Australia in 1955 to go ahead with the permanent atomic test site at Maralinga seems to have been predicated on Australia building its own atomic capability.[69] Australia looked seriously at acquiring atomic weapons, with encouragement from physicist Mark Oliphant, who 'advised officers from the department of external affairs that Australia should not agree to any proposal that would prohibit the manufacture of atomic weapons'.[70] To date, though, Australia has not been nuclear armed or powered. The politics has never quite worked out.

Lindemann often assisted in drafting replies to questions by backbenchers in the House of Commons, to refine and state the official line. One draft in the Lindemann archive dated 10 November 1953, just weeks after Operation Totem, dealt with Lindemann's unsuccessful visit to Australia to negotiate for the purchase of Australian uranium. Labour backbencher Maurice Edelman asked in parliament whether the mission had been a success and whether 'Her Majesty's Government can rely on future supplies of Australian and other Commonwealth uranium for the production of industry power from atomic energy; and how progress in the industrial atomic field compares with that in the United States of America and the Union of Soviet Socialist Republics'.[71] Lindemann's draft reply stated:

He [Lindemann] saw for himself the arrangements made for obtaining scientific data from the atomic weapons tests. He also ... obtained first hand information about uranium deposits and production in Australia and had discussions with Australian Ministers and with the South Australian Premier on Commonwealth and State programs for exploration and subsequent development. On his return journey through the United States he had discussions, in so far as this was possible within American law, about atomic energy matters of common interest. The military and industrial uses of atomic energy are so closely linked that it would not be in the public interest to answer the third part of the question. As to the last part, Her Majesty's Government's information about progress in the U.S.A. is insufficient to permit any comparison of relative progress in the industrial atomic field. Her Majesty's Government have no information about the programme of the U.S.S.R.[72]

After Lindemann had left Australia, Menzies wrote to Churchill to continue hedging on uranium supplies while praising the British prime minister's friend to the skies. 'The "Prof" is, of course, a retiring man, but his personal impact upon my colleagues could not have been better. He saw a great deal. In his discussions with us he was quite realistic. We learned a great deal from him.'[73] There followed a big 'but', after all the praise. While export of uranium for military uses was locked in via the joint UK–US CDA, civilian use of uranium was not – yet. 'But we were able to assure him that when we reached the point of uranium production for civil purposes we will, of course, regard the United Kingdom as a preferred customer in respect of any exportable surplus.'[74]

The following year, after the Rum Jungle and Mary Kathleen mines swung into action, the Americans also sent a delegation to Australia, as part of a new push to encourage civilian nuclear energy, to inspect uranium mines and discuss 'peacetime atomic development'.[75] The delegation reported that they were 'greatly impressed with the Australian atomic program, particularly as regards uranium ore development and processing in which fields Australia will play an increasingly important role both for defense of the free world and for peacetime development of atomic energy'.[76] The visit was not just about doing business with Australia, but was part of a bigger picture. The Americans had embarked on a major project after World War II to dominate both nuclear weapons and energy, and used its considerable strength in pursuit of that aim. It had reservations about Britain's attempt to match its endeavours. By the mid-1950s, with Calder Hall coming online as the first commercial nuclear power reactor, Britain appeared to be pulling ahead, not in nuclear weaponry but certainly in commercialising nuclear power.[77] In addition, Lindemann's brainchild, the United Kingdom Atomic Energy Authority, now separated from the civil service, was building up its strength.

President Eisenhower had promised that atomic energy would provide 'electricity too cheap to meter' and put forth an ambitious atomic agenda known as 'Atoms for Peace', which had a strong propaganda element.[78] He laid out this agenda in a speech to the United Nations General Assembly in December 1953.[79] This attempt by Eisenhower to diffuse fears about a nuclear future had the effect of increasing nuclear proliferation. Once countries decided to develop nuclear energy, they often also decided to obtain nuclear weapons.[80]

Shortly after Eisenhower's 1953 speech to the General Assembly, Australia was asked to join an exclusive group of eight nations to help draw up a statute for America's proposal for the formation of the International Atomic Energy Agency under the auspices of the United Nations.[81] This organisation began operations in 1957, although its original role as an agency to provide a bank of fissile materials for countries pursuing nuclear energy was abandoned before it came into being.

These early dalliances with the Americans in matters nuclear created ongoing relationships that led to, among other things, Australia bowing to pressure from the United States to sign the Non-Proliferation Treaty in 1970, and its commitment of the Joint Defence Facility at Pine Gap near Alice Springs.[82] In 1971, Rum Jungle closed down, having failed to live up to the overblown promises of the 1950s.[83]

An editorial in the Rockhampton *Morning Bulletin* summed up the extraordinary turn of events that gave Australia some bargaining power:

> It is strange that a discovery which has done more to give Australia additional world status than anything else that can be named offhand has become the subject of a rowdy domestic political squabble. Australia finds itself courted by governments, visited by distinguished scientists, caught up in a stream of great world events by reason of the discovery of uranium over a wide tract of territory that does not exhaust the possibility of further finds. Almost overnight we have emerged as a prospective major supplier of the precious and coveted ore, source of military power in the present and of Industrial power in the not distant future.[84]

Australia shifted, if only temporarily, from a small, marginalised country not long out of its colonial past, to a nation that had something the great powers wanted. Australia did little to place itself in that position – the key decisions and historical twists and turns were mostly external. But in the early 1950s, Australia could say yes to being the testing ground for the United Kingdom's nuclear weapons, building upon its existing agreement for the British to test rocket technology at Woomera. It could negotiate markets for uranium, one of the post-war era's most valuable materials: 'As gold was to colonial Australia, uranium was to the developing postwar nation', notes historian Robin Gerster.[85] Australia did at first harbour 'grandiose dreamings' about its uranium reserves, dreams that were never quite realised and the pursuit of which became politically toxic over the coming decades.[86] Just for a brief moment, though, as Britain tested its Totem bombs, Australia could see an atomic future dawning. It turned out to be a false dawn.

Chapter 10
MARIE CELESTE

It was very obvious that the place had been deserted in a hurry. There were private letters and wrist watches on bedside tables. It really looked as if an officer had just said 'get into a vehicle' and people had left. There were half full glasses of water with a little covering of dust on the surface of the water.

> Peace Officer Noel Williams, stationed at Emu Field in the weeks after Operation Totem.

[There are] places where sand had drifted and covered over small piles of condensers, resistors and other pieces of equipment.

> JF Hill, Report of enquiry into souveniring of official stores from Emu Field.

[A] lot of stores and equipment were loosely dealt with, and opportunities were not lacking for 'souveniring' on the part of Australian Service personnel.

> CA Stinson, acting secretary, Department of Supply.

Tents with dusty tables and even dustier glasses and tea cups, cutlery and plates both clean and used attracting flies, jumbled heaps of electronic equipment and snaking cables: X200 was a strange and eerie mess, without a human being in sight. The sudden abandonment of Emu Field, after the immense build-up to its short atomic test series, adds to its air of mystery. While there is no evidence that the site was deserted hurriedly because the village was accidentally contaminated by Totem II, the rumour lingers. John Symonds notes in his comprehensive 1984 account of the British tests that the winds at the time of Totem II shifted the ground contamination well away from the village, although some fallout did drop onto the roads.[1] Historians Lorna Arnold and Mark Smith reflect that '[f]allout had contaminated an unmanned photographic site on the eastern side of Emu Field, and this fact may have started the rumour that the camp was abandoned and never used again because of contamination'.[2] More likely the site was abandoned because it had fulfilled its function. Penney and co. were focused on the new and much bigger South Australian test site, soon to be named Maralinga. The scientists also had a swag of new atomic weapons data to work on back at Aldermaston. The flies and heat of Emu were uncomfortable. So the British test authorities moved on without a backward glance, and left their things behind.

The seeds of the demise of the Emu site were sown from the beginning. The RAAF historian Ivan Southwell said in his book on Woomera: 'They quickly learnt at Emu that Woomera had survived for one reason alone – it was served by its own railhead. Air lifts were possible – Berlin had proved it – but the cost was so great that it could be borne only in an emergency'.[3] Emu had the advantage of being extremely remote for secrecy, but its usefulness was outweighed by its difficulties. As the UK scientists turned their energies to design the components of a powerful fusion weapon, they looked elsewhere – Monte Bello again for Operation Mosaic (to meet urgent testing requirements, despite its logistical difficulties), Maralinga for operations Buffalo and Antler, and finally the Pacific for tests of the British H-bomb. A remote site with few options for access was no longer in the frame. The Royal Commission found 'no evidence to support allegations of an emergency evacuation of Emu after the Totem series. Stores which remained at Emu were low priority items not required in the UK or at Woomera.'[4] The desert, and the light fingers of some of the departing troops, soon claimed whatever was left.

In November 1953, Supply Minister Howard Beale wrote to Prime Minister Menzies to confirm that Emu was not suitable for future British atomic tests. This fact was probably already well known among the British authorities, but less clear to the Australians, who didn't always know what the Brits were thinking and were certainly not insiders on any decision-making. Beale wrote: 'On the question of a "permanent" site, Emu Field seems to be out of the question, mainly through shortage of water and difficulty of access.'[5] However, Beale also noted that the future of Emu was still uncertain, depending on which way the UK authorities wished to go:

> It is of the utmost importance that my Department be informed as early as possible if the next series of tests are planned within one year. If this be the case, there is no alternative to using Emu Field again, and plans for evacuation of stores and equipment will have to be modified.[6]

That was not to be; the next British atomic bomb on Australian territory was detonated back at Monte Bello, two and a half years after Totem.

The media speculated a little about the expense involved in setting up a site that was so quickly abandoned. The Adelaide *Advertiser* ran a story in November 1953 that suggested Emu would no longer be used:

> The establishment of Emu Field testing base was a costly project because of the huge transport costs involved. The British organisers of the atomic tests are understood to have reached the conclusion that it would be better to 'cut the losses' on Emu Field by abandoning it.[7]

The story suggested that a new site nearer to the east–west rail line was being considered, as indeed it was.

After the second Totem test, the scientists and most of the other on-site crew departed immediately, with a team of 22 left behind to wind things up.[8] This group was required to handle the evacuation of remaining stores by road 'and to enable this area to be closed down in a manner that will provide maximum protection for buildings and equipment remaining at site'.[9] The plan was to keep some stores at the site, just in case the British test authorities decided to return. This included all cooking, messing and refrigeration equipment, and a couple of generators. The huge British-built Centurion tank that was ferried to the site in July was laboriously moved to Woomera, from where it went on to Puckapunyal and thence to Vietnam as the legendary 'atomic tank'.[10] The tank, known to troops in Vietnam as Sweet Fanny, had been purchased under contract in 1952 by the Australian Government for more than £20 000.[11]

Emu Field atomic test site was evacuated for good, other

than its Peace Officer Guard and the last remaining military personnel, on 29 November 1953, just over a month after Totem II.[12] The departure did not always go smoothly. The final stage of the evacuation was arranged as a land convoy, with vehicles having to negotiate harsh terrain. About 40 kilometres north of Kingoonya, around midday on 1 December, a fire broke out on the rear end of one of the Land Rovers. The vehicle was carrying personal belongings of five RAAF personnel. At the time, the vehicles were travelling according to the orders provided to the men, at intervals of approximately 0.8 kilometres. The driver of this particular vehicle was alone because of a shortage of personnel. He noticed smoke coming from the back of his vehicle, so he stopped and used a fire extinguisher, but it was too late – the auxiliary petrol tank caught fire and soon it was too dangerous to approach the vehicle. All the belongings of the RAAF personnel were destroyed, and the vehicle had to be towed.[13]

Because the men who lost their belongings were seconded to the Department of Supply for their construction work at X200, they were regarded as civilian employees of the department. An investigation showed that while the exact cause of the fire could not be identified, the Land Rover had no muffler on its exhaust system owing to an acute shortage of spare parts at Emu. The convoy was driving in extremely hot conditions in difficult country, and it was speculated that the heat from the open exhaust pipe sparked a fire in nearby combustible material. Since the men 'were compelled to travel and work on this project', compensation for the loss of their belongings was recommended, for a total of over £114 between the five of them.[14] Among the burned belongings were cigarettes and tobacco that were part of the men's five-month leave ration.[15] One of the men was a stamp collector, and he lost a lot of his new stamps. Another man put in a separate claim for £10 worth of opals. Small losses, but significant to the individuals concerned.

While the final items of gear were moved across country, more was taken out by air. During the evacuation period, two daily Bristol flights were scheduled for a week to airlift supplies. They were working to a hastily arranged protocol. A report from the Australian Attorney-General's Department set out the requirements for the evacuation. A Department of Supply report outlined them:

> Immediately following the second explosion, it became evident that the United Kingdom staff were anxious to return to the United Kingdom with the least possible delay, and stores and scientific equipment were packed ready for dispatch from the site as quickly as possible. These were graded into five categories:
>
> A. To be returned at any cost to the United Kingdom as soon as possible.
> B. To be returned at any cost to the United Kingdom, but with less urgency.
> C. To be returned to the United Kingdom because they are economically worthwhile.
> D. To be stored at Woomera or Salisbury.
> E. To be stored on site against the possibility of use in future trials.[16]

'The Australian personnel', the report continued, 'were also anxious to vacate the site with the minimum of delay and, after ten months without leave at Emu, this viewpoint can be appreciated'.[17] Emu was a hardship post. The men stationed there faced difficult physical conditions and onerous secrecy requirements, and their safety was not well looked after. They were typically in the dark about the unseen radioactive dangers of the place and the risks they took home. Back home they could not properly discuss their experiences with their families. Many kept silent about being there for the rest of their lives, and their service records showed no indication of either X200 or Emu Field. Their year of living dangerously and secretly was consigned to the past. Too many later developed significant health issues that may have been connected to their service at Emu.

By 8 December, with the main evacuation complete, only eight men remained at X200. The 'problem of maintaining them' fell to the Long Range Weapons Establishment at Woomera, which planned a flight to Emu every Wednesday, 'calling at Ooldea if necessary', to send supplies to and collect mail and other items from the men still there.[18] That weekly flight from the RAAF base at Mallala via Woomera must have been a highlight.

Radioactive contamination forced the Department of Supply to write off a number of Commonwealth vehicles, including four Land Rovers, three trailers and a Jeep: 'it is not practicable to decontaminate them and make them safe for use again'.[19] The complications of decontamination procedures in the desert were simply impossible. 'These vehicles were required to perform such functions that they could not be protected from the intense radio activity and their loss, therefore, must be expected as a necessary part of the experiment.'[20] They were to be buried at the site, contamination and all.

The tainted site was soon derelict. While the rapid departure of scientific staff was understandable, given the urgency of weapons development and the mountain of new work for them back at Aldermaston, the complete and sudden abandonment by everyone involved inevitably caused speculation. The Marie Celeste vibe of the departure is palpable, down to the dusty glasses of water on the mess tables. The Australian Department of Supply endeavoured to follow its orders to get out of Emu quickly.

Was it a misunderstanding, or simple thievery? An Australian serviceman was court martialled for stealing materials from Emu Field, and Minister for Supply Howard Beale was asked questions about the charges in parliament. Beale initially said that nothing of value had been stolen, which he later had to clarify, assuring the house that the items taken did not compromise the security of the site: 'I did not say that the articles taken were of little value. I said that they were not of security value'.[21] The serviceman involved, Corporal Desmond Francis, claimed that the British scientific personnel had told him to take whatever he wanted. This was later denied by the scientific officer-in-charge, Charles Adams, and a he-said-he-said dispute unfolded.[22] Just what happened out of sight in the desert?

In his statement, which was attached to Hill's report, Corporal Francis said that he had arrived to undertake 'Special Duties" at X200 in March 1953. When his commanding officer, Lieutenant Olney, returned to Woomera, Francis was left in command of signals at Emu, with instructions to install communications equipment and make contact with Woomera.[23] After the second Totem test, work at X200 wound down and the men relaxed for the first time since they had arrived. The signals operators, among the last to leave, had more time than others to pack their belongings. At the invitation of a scientific officer, Corporal Francis visited the test area.

A lot of gear was being burnt and thrown out; I was told I could have anything thrown out of the gear strewn around the half empty and broken crates, I could take my pick ... I picked what I wanted; a lot of the components would have been covered by the shifting sand and never recovered if I had not found them – no doubt there is a lot I never found. Scientific Officers and/or Officers of the Army and R.A.A.F. were present when I collected most of the components – they did not reprimand me in any way. I took all the gear to the transmitter shack, parcelled it, took it to the orderly room where it was weighed, postage paid and sent on its way via the aircraft.[24]

Francis said he was 'given a free rein in a paradise of radio components'. He planned to build himself a workshop back home and fill it with equipment, for his own enjoyment. As well as the gear he posted, he took many radio components with him when he left by road. Much to his surprise, he was apprehended by a security officer the day after he rolled into Woomera. The gear he had collected was stacked up in front of his hut, unconcealed.[25]

Francis had been caught red-handed, having drawn attention to himself by posting such a large volume of items. This alerted the government that something was amiss.[26] In January 1954 the Department of Supply asked JF Hill, an investigator, to enquire into the 'alleged souveniring' of equipment from X200 by Australian Army personnel.[27] He delivered his report to the Commonwealth Investigation Service, part of the Attorney-General's Department, in April. Hill made a point of saying that he had ignored everything he had read in the papers about the incident and relied instead on evidence and a visit to the site.[28] It whipped up a brief flurry of media interest that soon blew away.

Until May 1953, security officers at X200 subjected all outgoing parcels to external security. Head of X200 construction, Brigadier Lucas, stopped the practice at the beginning of June because it slowed down mail deliveries.[29] Lucas was advised that paint brushes and bed sheets were being mailed from the site, apparently quite blatantly, although no action was taken. Hill suggested turning a blind eye helped to keep up morale in the face of harsh working conditions:

> The Brigadier ... undoubtedly realised that by making one man a defaulter out of so trivial a charge would cause dissension, and that he could not expect the co-operation he was receiving from the men to continue, considering also the time that would be wasted conducting an enquiry on the spot, and the consequent action, awarding of punishment and performing of defaulters drills by the accused, would all have tended to slow down the pace of the preparation work.[30]

X200 was a rough and ready frontier settlement, and frontier standards applied.

Once the UK party started to arrive in August 1953, the commanding officer sought to foster close working relationships between the Australian and UK personnel. This admirable ambition ultimately led to the downfall of Corporal Francis. Soon after the tests had finished, the UK party packed up and left the site, leaving instructions about how the remaining supplies were to be handled. The Australians still left at the site parcelled up large quantities of their own possessions and, it turns out, other materials from the site, and sent them by post.

For a while, about 140 parcels per day were being received at Woomera from X200, flown by air courier to the Woomera post office.[31] Mostly the parcels consisted of uniforms and personal equipment, as well as hand-made ornaments sculpted out of mulga wood, one of the main downtime activities at the site.

The final convoy, including Corporal Francis, left Emu on 3 December, taking with them towels, overalls, a reading lamp and tools. 'Their total spoils would probably [weigh 152–200 kilograms]; comparing this with the [2267 tonnes] of stores and equipment that went to X200, the percentage is small', Hill found.[32] He also discovered that UK staff had discarded pieces of electrical equipment – including condensers and resistors – in the various X200 rubbish tips: 'In almost every hut I saw articles of small value which did not appear to have been used, yet were discarded, and which even now may have deteriorated through weather conditions to such an extent as to make them unusable'.[33]

Hill concluded that any future project of this kind must impose strict controls on items posted from the site:

> Had there been censorship of mails on this project, the question of souveniring would not have arisen. A kit inspection of all service personnel should have been carried out before entering the area and also on leaving; this included in the drill would have given satisfaction to the men and to the authorities.[34]

The items recovered from 34 Squadron at RAAF base Mallala included a Breezaire air conditioning unit, 14 tubular steel canvas chairs, two cane chairs, two tables, a hydraulic jack, a kerosene heater and an electric motor.[35] Almost certainly these items would have been left to rot in the desert if they had not been collected and shipped out by a few military personnel.

Hill concluded that the incident was magnified beyond its importance, as was the value of the recovered items. Most of the items were defective anyway, he said, which is why the UK scientists had discarded them. He also stated that:

> The area X200 appears to be in need of a general cleaning up. But the point first to be decided is would the cost of labour justify the results. 'E' site in particular, is in a dirty condition – it gives one the impression of having been hastily evacuated. Dirty plates, glasses, knives and forks, un-made beds, are still in the huts just lying where they were left; the Q.M. [quartermaster] stores has [sic] a sizeable quantity of kitchen and dining-room equipment, including crockery etc. that would have to be packed before it could be moved.[36]

He did not blame the men who were the last to leave the site, since all at X200 had contributed to the fact that the packing and protection of stores was never carried out properly.

When Hill interviewed Major Lowery in March 1954 about the souveniring, Lowery, who was in charge of the X200 construction crew, told him that the most coveted items on-site were towels:

> After the men had been working in contaminated areas, the towels which were used after the men had had a shower were often thrown into a pit and covered over. When the troops saw this, they apparently decided to use Army issue towel (which is of poorer quality) throw it away after their bath and keep the new one as part of their kit.[37]

Such small lurks and perks no doubt occur during many military deployments.

In an appendix to his report, Hill included a personal, secret letter from Dr DH Black, head of the United Kingdom Ministry of Supply Staff Australia, who wrote to Captain FB Lloyd at the Ministry of Supply in December 1953. Black said

> we both feel that the matter should be hushed up as much as possible. We feel that if the man concerned is courtmartialled, there will not only be a public stir about the matter, but that a good counsel on his behalf could throw a considerable amount of mud.[38]

Captain Lloyd responded, 'I am inclined to agree with you that we don't want a lot of mud-slinging and that a reprimand by the man's C.O. might prove the best answer.'[39]

The souveniring problem made it into the media. The Adelaide *Advertiser* reported that the official explanation for the disappearance of 'secret equipment' was '[s]ouveniring run wild':

> When the Australian troops were helping British scientists dismantle observation towers, they are believed to have been told that some of the gear was not worth taking back to England. The surplus gear included batteries and small radio sets. The troops are reported to have interpreted the permission liberally, and their 'souvenirs' included some secret electronic equipment. It was stressed ... that no Australian equipment had been 'lifted'. The authorities are puzzled what action can be taken so long as the men

adhere to their story that they were given permission to take the gear.[40]

Unable to work out exactly who had said what to whom, the senior officers simply ignored most of the souveniring. No doubt many suburban sheds contained small items from X200, with a proportion of them a little radioactive.

Notwithstanding trivial pilfering, security was an expensive and ongoing matter at X200. At the height of operations at Emu, nine peace officers maintained security.[41] Once the site began to wind down, these officers were withdrawn. The appointment of peace officer Frank Moffitt at the end of October 1953 marked a change in the civilian security operations at the site. Unlike other peace officers, Moffitt was trained at Woomera specifically for this task by scientists who had been posted at Emu. He was required to understand the Commonwealth Crimes Act and several other highly restrictive Acts that related to Woomera weapons testing.[42] Moffitt served in both world wars, and was familiar with the Woomera area. He was an expert in bomb disposal, and had trained troops in these skills during World War II. His appointment was endorsed by a senior peace officer, E Hattam, who commented: 'With his background and training on scorched earth policy, I feel that he has the requisite experience and qualifications to perform the task to be undertaken at X.200 site'.[43]

Moffitt was aged 60 when he was assigned to work seven days per week at Emu, a site renowned for its harshness.[44] Sergeant Newall, who had been there since June, initially remained at the site in order to assist him 'as it was felt that with the duties involving attending to instruments, vehicles, etc. the duties could not be wholly performed by one officer'.[45] Moffitt was joined by two other peace officers when the main maintenance party withdrew from the area, although at the start of his posting he worked only with Newall. He kept himself somewhat aloof from the other peace officers. Noel Williams, who stayed for about a month from mid-November to mid-December, did not take to Moffitt. He found him a rather distant, secretive and unfriendly person who had 'just appeared out of the blue and was wearing three stripes'.[46] The men were supplied weekly by air, and provided with fresh and tinned food sufficient for three men for six months, as well as both fresh and salt water, an ablution hut and a 'Tumbling Tommy' latrine, three vehicles with fuel and spare parts, a power generator, communications gear, tented accommodation, a first aid kit and medical supplies, and the all-important darts board and table tennis equipment.[47] Their lonely post was hard work but had some small comforts.

The Deputy Superintending Peace Officer Hattam wrote to the superintending peace officer just before the tests saying that he had been briefed that 'the area will remain contaminated and radio active for a period of perhaps up to forty years'.[48] After the tests, the Long Range Weapons Establishment (LRWE) at Woomera had the authority for range security. The LRWE security officer ordered peace officers stationed at Emu to carry out 'radiological patrols' in the contaminated areas.[49] They were also told to take showers immediately after patrolling the bomb sites.[50] Peace officers wore film badges, although exactly how or when the badges were examined for evidence of exposure to radiation is not clear. CS Morrison, X200 project security officer, was given orders that '[l]arge parts of the area will remain contaminated for some years and will require both security and safety guard'.[51] Morrison advised the LRWE that peace officers at the site were required to treat as classified a variety of information about the site, including:

(a) Size and extent of craters.
(b) Alpha, beta and gamma radiation readings in the contaminated area.
(c) Samples of radio active soil, flora and scientific exhibits, etc.
(d) Extent of blast areas and distances between observation posts, camera posts and explosion areas.
(e) RH scientific instruments and equipment (unclassified but of high monetary value and hard to replace).[52]

One of the main roles for the peace officers was to enter the areas where the explosions had taken place, to look after the instruments that remained in place and change the batteries in the machines. Williams gave evidence at the Royal Commission that he had been provided with a crude, pencil-drawn map that indicated the contaminated areas at the Emu site. Moffitt showed him a wire and steel fence that was said to be contaminated on one side but safe on the other, instructions that provoked some understandable scepticism.[53] When the peace officers came out of the contaminated areas they would check each other for contamination in a marquee near the test sites. Williams said that he was found to be contaminated wherever perspiration was on his body, in the neck area, under his arms and on his feet. He would shower in cold water and a kind of detergent called Teepol.[54] He commented ruefully, '[o]n reflection, the officers were not stupid because at no stage did any officer go into the [contaminated] area. They always sent men like myself'.[55] Curiously, Williams was also asked to look for 'confidential papers … These are papers which should have been destroyed. We spent a bit of time looking for them but never found any'.[56] He described some typical activities he undertook only weeks after the atomic bombs were detonated:

> We would go into the dirty area early in the morning about 10 o'clock and would come out later in the afternoon, at about 3 o'clock. The area was very bare and barren and was very prone to dust and sand storms. The respirators used to get literally filled up with sweat so that there was a real danger of drowning. We would have to pull the respirators up quickly and allow the perspiration to drain out. Additionally, the glasses would fog up. I used to carry a rag up my sleeve and I would quickly turn them up and clean them when they got so bad that I couldn't see out of them. I think that I kept this rag in my sleeve or my glove and I believe that I used to wash it out when I got back.[57]

Williams later developed emphysema and problems with his hands (possibly from climbing observation towers that had been contaminated), and his wife suffered repeated miscarriages.[58]

One peace officer, Thomas Murray, stationed at Emu after Totem, took it upon himself to go searching for dingoes with a view to killing them.

> I had a habit, and it was deliberate, that I walked many miles for the purpose of – to see if there was any native camps, any native tracks, and also I had dingoes in view – tracks of dingoes in view too, because they were a nuisance as far as the pastoralist people were concerned and – from a very young person, always destroy a dingo because they destroy many sheep.[59]

It appears from his evidence that the local dingoes managed to evade him.

By April 1954, the need for regular reports from X200 ceased and the range security officer, EA Hayes, suggested that only the senior peace officer Moffitt remain on the site.[60] Eventually, in 1955, Moffitt left, although peace officers including Thomas Murray still visited Emu about once a fortnight, travelling up from his new base at Maralinga on the goat track of a road. Murray ended up with blood and bone marrow disorders after his service at Emu and Maralinga, and said in his statement to the Royal Commission '[t]he preventative measures were inadequate', a regular refrain from those exposed to radiation from the tests.[61]

Murray also remarked on a strange aspect of the areas of the atomic tests: 'The animal life disappeared. I was an experienced bushman which was the reason my services were sought and I noted that change in the whole area as a result of atom bomb testing'.[62] The absence of animal life even now is a curious phenomenon.

A military installation created at huge expense started to be covered over by the relentless desert. Given the cost of establishing the Emu site, the cavalier abandonment of it seems extraordinary. When Frederick Lindemann had first approached Prime Minister Menzies about using X200, he had indicated that the UK government would reimburse the Australian Government for the costs of setting up the site. A formula was worked out to apportion the costs. Australia was to cover the salary costs of the Australian public servants involved in the tests, the pay and allowances of Australian service personnel and the cost of transport within Australia of UK personnel. In addition, Australia would provide all the site's motor vehicles, campsite gear such as tents and cooking equipment, signal communications between Emu and Woomera and medical facilities in the operational area.[63] The United Kingdom would be responsible for the overseas allowances and overseas travel expenses of Australian public servants working on the project, depreciation on recoverable materials and the cost of materials that were not recoverable, the maintenance and operational costs of vehicles and aircraft including the cost of any vehicle or aircraft destroyed by the operation, and all costs associated with UK personnel.[64]

At the conclusion of the Emu tests, the Australian Government was out of pocket by over £828 000, or approximately $30.2 million in today's money, about £30 000 above the original estimate by the Prime Minister's Department.[65] The government sought to bill the United Kingdom for just under £700 000 of that cost. There was considerable confusion about just how much Australia could get back, particularly as the Australian Government did not know at that point whether the United Kingdom would continue testing at either Emu or somewhere else. In particular, the £115 201 for stores and supplies, £53 859 for building materials and £237 197 for plant and machinery were grey areas. The secretary of the Department of Supply, Frank O'Connor, wrote to Treasury in February 1954:

> Many of these items are still in position at the atomic bomb site, and to enable an accurate division to be made, it will be necessary to determine a residual value under each classification. Until a decision has been reached regarding the possibility of a future permanent atomic test site, it is not possible to allocate, even on an arbitrary basis, the costs between U.K. and Australia.[66]

Earlier, in October 1953 in the House of Commons, the UK minister for supply, Sandys, quoted in the journal *Nature*, had stated that his government should bear the future cost of any further atomic testing in Australia.[67] In the event, Australia paid a considerable amount of money towards the later testing at Maralinga as well. The country sank a lot of money into test sites that became headaches decades later and would one day cost millions of dollars to remediate. As the memory of what went on there quickly faded, not too many questions were asked about the value Australia had received for its significant investment.

Chapter 11
THE FORGETTING OF EMU

```
The three months of the operation have made little
impression on the mulga. The rabbits will do more
damage in a day than we have done in all this
time.

        Voiceover, official UK AWRE film of Operation Totem.


Many are still suffering today. The emotional,
mental and physical suffering is felt by
generations ... We are constantly reminded of
what has been taken away from us as a family
and the suffering we have gone through.

       Karina Lester, Yankunytjatjara-Anangu woman,
    daughter of Yami Lester, speech to the 2017 United
    Nations Treaty on the Prohibition of Nuclear Weapons
                                         negotiations.
```

Emu Field is ethereal, like a ghost or a faint echo. This far-off place did not enter the national psyche in the way Maralinga did (albeit after many years). While Maralinga is a familiar name, Emu Field is not. The legacy of Emu is largely unknown, and the British government still keeps its Emu secrets suspiciously close. Getting access to British files on Emu Field has always been more difficult than getting Maralinga files, although both are highly restricted. Even before the mass withdrawal of British files in December 2018, Emu records in whole or part were often closed to scrutiny, retained by the UK Ministry of Defence. In some cases, material that was once available was withdrawn during the 1990s. We are left to wonder why.

Some things we know for sure. The British, without hesitation, turned a blind eye to the fact that the Emu Field site was close to human habitation. The official line at the time was that protecting Aboriginal people was Australia's responsibility, and Britain has maintained that stance to the present day. Some British documents did allude to 'difficulties about Australian Aborigines', but these issues were never pursued.[1] The country took the same approach to hydrogen weapon testing in the Pacific in the late 1950s, when Pacific Islanders were put in harm's way without any apparent discomfort on Britain's part.[2] That Australia largely failed in its undeniable role in keeping Aboriginal people safe does not absolve the British. Their priorities during 11 years of atomic testing in Australia were for the rapid development of an effective, deployable and cheap atomic weapon, and after 1954 the fission components of an H-bomb. The British authorities did not always hold the safety of its own military and scientific personnel in high regard; the lower priority it accorded Aboriginal people is beneath contempt.

The passage of time caused some changes in attitudes. For example, the architect of the British bomb, William Penney, became a critic of nuclear testing, and in the 1960s he detached himself altogether from Aldermaston. While his resolve during the early days was rock-solid, it appears to have shifted as the worldwide ramifications of nuclear testing became impossible to ignore. Penney was a rather introspective person who did not fully exonerate himself from unleashing the atomic genie, and he attempted to make amends in his later years. Given his role as part of the inner circle of the Manhattan Project that created the bomb, the problems with his conscience are understandable and indeed were shared by many physicists, some of whom came to believe that their scientific discipline was tainted by the bomb. Penney was thought by some to be 'the British Oppenheimer', referencing the scientific leader of the Manhattan Project, Robert Oppenheimer, and the comparison is not entirely unwarranted.[3] Churchill and Lindemann thought him as brilliant as Oppenheimer, but more reliable and less politically risky than the left-wing American. Penney was always more comfortable behind the scenes, sleeves rolled up, brow furrowed, working on a complex nuclear conundrum and, it seems at least possible, not looking too far into the real-life implications of solving those intellectually stimulating riddles and creating a weapon of war.

Ultimately, he was moved by those consequences and the unease they caused him:

> Behind the scenes, he worked on negotiations to ban nuclear tests, but was disappointed when they ended in 1963 with an agreement to outlaw only atmospheric tests, leaving open the option of underground detonations. He retired from the nuclear industry in 1967 and accepted the post of Rector of Imperial College.[4]

Penney watched on as stockpiles of these weapons grew around the world. As author Graham Farmelo wrote:

> In private, Penney was sometimes scathing about the way governments had handled nuclear weapons. When British

historian Lorna Arnold asked him why the superpowers had stockpiled more of them than they could possibly use, he replied, 'Because they were mad, mad, MAD!'[5]

Like many intellectuals of the time, Penney was drawn to the anti-nuclear movements that sprang up in the late 1950s. He was in good company. Farmelo remarked:

> To the surprise of some of his colleagues, Penney joined Pugwash, an organisation that brought together leading scientists to discuss how to minimise the threat of nuclear war, with the aim of providing apolitical advice to governments ... Its attendees included Mark Oliphant and Leo Szilard.[6]

Penney also came to distrust journalists and destroyed all his correspondence in later life, much to the despair of historians (especially since the British Government end of his more interesting correspondence is unavailable, deeply buried in the archives at Kew).

He had become so removed from his old life as the destroyer of worlds that by the time he appeared before the McClelland Royal Commission in London in January 1985, he was blindsided by the rigour and tenacity of the questioning and not always able to control his usually slow temper. The Royal Commission got under his skin, possibly because feelings he thought had been safely buried under layers of protective concrete were so easily dredged up again. He helped to let the atomic demon loose in the world and had to find ways to live with that fact.

Penney was a naturally polite and easy-going chap, one of the reasons he was well liked. He sent a warm letter to head of the Australian Totem Panel, Jack Stevens, soon after Operation

Totem, telling him that the Totem tests had 'been completed with gratifying technical success, and without any untoward incident', as ever brushing over the more disturbing details:

> When the decision was taken in the United Kingdom early this year to request Australia for permission and help in running the Atomic Tests this year, we realised how big was the effort we were inviting Australia to make; but the job was accepted and then well done. I am proud to have worked with the men you sent to Emu. Australia has never chosen a better Test team![7]

Penney was always fond of a cricketing analogy. But his likeability and evidence of a conscience do not excuse his casual dismissal of safety issues:

> As a result of the Totem experience, I can state with confidence that if Atomic Trials are ever again held in the Emu area, fall-out of radioactive dust on the Mainland of Australia need not give cause for the slightest anxiety, provided the Scientific Director waits, as of course he will, until the winds are suitable.[8]

Was this sheer deluded optimism, or was Penney a loyal servant of a government with a vested interest in calming fears and providing specious assurances? He went on:

> Because of the rapid decrease of radioactivity with time after explosion, the activity in the Test areas is rapidly clearing, and in a few weeks' time the only dangers left will lie in the fused regions immediately surrounding the two explosions sites. Arrangements which have been made for

guarding these two small areas are ample to ensure that no ill effects will be experienced by any aboriginals who may enter at some time, from curiosity or in the hope of finding something useful.[9]

That strange, dark trinitite of the 'fused regions' remains scattered around the Emu blast sites like smashed up bottles.

Penney was generous in his praise for Australian personnel and always quick to show appreciation:

> While I would not wish to single out any of the Australians at Emu as standing head and shoulders above the others, for everybody played his part well, I feel that it is only fair to mention the senior officers, Brigadier LUCAS, Squadron Leader GARDEN and Major LOWERY, as having worked splendidly, without stint, on an exacting job.[10]

He also lauded the RAAF personnel who participated: 'The cheerful and impressive performance of the aircraft crews, among whom the Bristol crews were outstanding, made the Tests possible, and to them also is due the thanks of the United Kingdom.'[11] His praise extended to every member of the small Emu band:

> Finally, among the X200 force, must be mentioned the cooks. They never gave us a poor meal, and quite often they put on a magnificent spread. This was only achieved by long hours of cheerful work, in conditions which would have given them every excuse for shortcomings. The cooks kept us going and they deserve our best thanks.[12]

Towards the end of this long paean of praise, Penney said, 'The scientific contributions made by the Australian scientists, especially Professor TITTERTON, will add appreciably to our understanding of the neutron flux and spectrum in the damage area of atomic explosions'. In conclusion, he said:

> To you and your staff, to the Australian Totem Panel, and to all the H.Q. staff in Canberra, Sydney, Melbourne and Adelaide, again my best thanks, but worth more than these is the clear demonstration which has been given to the world of an ambitious technical operation successfully accomplished, and proving once again how well Australians and Britishers can work as a team.[13]

The team was lopsided in its power – one side knowing far more and having the monopoly on decision-making – yet there is no reason to doubt the sincerity of Penney's words. The sense of shared purpose comes through. But Penney walked away unscathed from the events he is talking about, with the privileges that come with colonialism. He left the aftermath of his work for others to deal with, sometimes decades later. The Australians who were in the line of fallout of Operation Totem could not be so sanguine about tests that were not well prepared for, monitored or reported.

Totem was not properly set up. The shortcomings were many. These were canvassed by the 1984 Kerr report, known officially as the Expert Committee on the Review of Data on Atmosphere Fallout Arising from British Nuclear Tests in Australia, chaired by Professor Charles Kerr. This report noted that that when the ATWSC reviewed the criteria for Hurricane and Totem in 1955, they concluded 'that meteorological facilities at the previous trials were inadequate'.[14] John Moroney and Keith Wise from the ARL noted in their extensive and well-crafted submission to the Royal Commission that there was no monitoring of the fallout across the country from the Hurricane and Totem tests.[15] When working out how much of the Australian population was exposed to fallout, they had to rely upon data from the later tests. And, Moroney and Wise readily admitted, '[l]ong-range fallout from [all 12 major atomic] tests reached both rural and urban centres and led to irradiation of the populations'.[16] Penney's air of bonhomie becomes less endearing in light of these disclosures.

Ernest Titterton noisily maintained until his death that the British nuclear tests in Australia had caused no harm to anyone. Even before the Royal Commission had released its report, he drafted a letter for the media which objected to its findings, and asked his former AWTSC colleagues to sign it. The draft letter said in part,

> the entire test series 1952–1963, although it caused exposure in parts of the population to low levels of ionising radiation, did not lead to any significant health effects. The investigations of the Royal Commission itself, have not revealed any case of injury to health, let alone deaths, either for the general population, white or aboriginal, or to persons directly involved in the programme. This confirms findings already made by the National Radiation Advisory Committee (1958–9), the AIRAC Report No. 9 (1983) and the more recent 'Kerr' Committee Report (1984), although this latter had some reservations for certain small groups.[17]

The AIRAC report was a long way from being the last word on the aftermath of the South Australian tests and was strongly criticised in both the Kerr and the Royal Commission reports.

Titterton contacted William Gibbs, director of the Commonwealth Bureau of Meteorology, who had served on the AWTSC from 1962, to become a co-signatory. Gibbs told him he did not want to add his name to the letter, noting that he and Titterton disagreed on the clean-up at Maralinga, while acknowledging his regard for him:

> The dangers to health from nuclear weapons tests will always be a matter of opinion as there are many matters physical and biological so there will always be debate on whether there was some biological damage as the result of the tests. I imagine that you would maintain that there is no question of such damage whatsoever. I would prefer to say that it is possible that some damage may have occurred in isolated instances, but that it is impossible to identify such damage because it would lie within the 'noise' of the health statistics of the people involved.[18]

Since no such health statistics existed for the populations most at risk, it was impossible to assess the harm.

Gibbs was also keen to place the blame elsewhere, including changing attitudes and the buck-stops-here responsibilities of government:

> The most important relevant fact is that the tests were carried out with the full knowledge and agreement of the Australian Government. There may be some debate on the wisdom of that decision but they were vastly different times with different attitudes from the present. I can remember, almost forty years ago to the day, wandering through the ruins of Hiroshima with never a thought given to possible hazards from radiation.[19]

Gibbs commiserated that 'Diamond Jim [McClelland] and his henchmen/women' would be tabling the Royal Commission report in parliament on the day he wrote his letter. 'I hope it does not give you too much concern. You should be secure in the knowledge that you did what you thought best for Australia.'[20]

Donald Stevens, another AWTSC colleague, also said no to Titterton's letter. The former head of the Commonwealth X-Ray Laboratory (later ARL, now ARPANSA) responded:

> I feel it would be inopportune for former AWTSC members as a group to comment now in the newspapers, sight unseen, on any of these submissions ... I am sure we will all read with interest the report of the Royal Commission. Predictably the report will raise controversy. In that regard, some months ago, I discussed informally with a widely experienced barrister-solicitor friend of mine the possible ways in which I should react to observations in the report of the Commission. His strongly expressed advice was for me not to make any media comment before I had read the report in full and especially to avoid fuelling or exacerbating, through media comment, any controversy that arises. I intend to follow that advice.[21]

Members of the old Australian safety committee correctly foresaw the fallout for themselves in the revelations of the Royal Commission. The safety committee was roundly criticised in the commission's report.

Titterton's letter was published by the *Canberra Times* on 30 November 1985, co-signed only by Alan Butement.[22] Titterton also wrote a letter of critique to the Royal Commission. His letter did not hold back. After outlining all the ways he had been accommodating to the requirements of the Royal Commission, in his trademark forward style Titterton said:

> At the Inquiry it soon became apparent that the Counsel assisting the Commission, Mr P. McClellan, had adopted the role of hatchet-man ... He was aggressive, and sought the worst possible interpretations of situations and documents, repeatedly indulging in all sorts of unpleasant innuendo. He pursued a line of leading questions in an attempt to support what was evidently a preconceived position, rather than to elucidate the truth of the events which occurred 25–30 years ago.[23]

Titterton was clearly wounded by the Royal Commission's process even before he read the personal criticism directed at him in its report.

The Royal Commission had been called amid growing anger at the cavalier approach of Britain to its responsibilities. Yami Lester had been infuriated by Titterton's suave assurances in a radio interview in 1980 and started his long advocacy for accountability. In addition, starting in the 1970s, whistle-blowers blew their whistles, investigative journalists asked impertinent questions, and politicians raised issues in federal parliament.

A flurry of interest in what the British left behind at their three Australian test sites led to several reports during Malcolm Fraser's prime ministership, and in the early stages of Bob Hawke's tenure. These two Australian governments, one Coalition and one Labor, gradually recognised that no Australian official actually knew what was at the sites. This was despite official but unreliable British documents, most particularly the 1968 Pearce report, authored by AWRE scientist Noah Pearce and later found to be completely wrong about contamination, especially at Maralinga. An AIRAC report tabled in federal parliament in 1979, titled 'Radiological Safety and Future Land Use at the Maralinga Atomic Weapons Test Range' contained a number of recommendations about erecting fences and warning signs at the Maralinga and Emu sites.[24] A report commissioned by AIRAC on Emu Field (Emu report) was completed in 1979 and presented to the Minister for Science and the Environment James Webster on 19 October that year. The report, which was tabled in parliament in 1980, was based on field survey work carried out in 1978 by members of the ARL, with assistance from the South Australian Health Commission.[25] The report suggested that levels of radiation around the two Emu ground zeroes were such that the areas would be unsafe for permanent occupation, although short visits would be acceptable.[26] Concrete plinths were erected at both Emu bomb sites to warn visitors that atomic tests had been held there and the ground zeroes were not safe for permanent occupation.[27] Still later, the infamous AIRAC 9 report of 1983 dealing with Maralinga and Emu failed to uncover the true extent of plutonium contamination at Maralinga.[28]

The Emu report found the worst damage was confined to a limited area: '[s]ignificant radiation levels due to long-lived neutron activation products in soil [cobalt-60 and europium-152] occur only in the immediate vicinity of the ground zeros of TOTEM 1 and TOTEM 2'.[29] These radioactive products of fission explosions were formed when the soil beneath the towers was irradiated with neutrons from the explosion.[30] In addition, the site had particles of unfissioned plutonium, the most dangerous substance of all. The scientists who undertook this survey in November 1978 noted in their report, '[b]etween a half and one kilometre from the ground zeroes there remain the skeletons of trees killed at the time of the explosions and, beyond about one kilometre, there are living trees up to six metres or so in height'.[31] The scientists said that the 'only significant path' for visitors to ingest radioactive materials at the site was by inhaling them. The Totem II site had five times the allowable maximum for continuous exposure and nearly three times as much airborne plutonium as the Totem I site.[32] These sites, left as they were in 1953, were not entirely safe in the late 1970s. The contamination at Maralinga was even more significant.

Subsequent analyses include a 2013 study by scientists DP Child and MAC Hotchkis, who reported measurements of plutonium concentration and isotope ratios in the soils of Emu Field (as well as Monte Bello). They used accelerator mass spectrometry with samples ARL collected in field surveys conducted in 1972 and 1978 – the same samples used as the basis of the Emu report.[33] The origin of the plutonium measured in this study was taken from residual unreacted weapon material vaporised during the explosion and deposited from the atomic cloud and the fallout plume, the same unfissioned plutonium noted in the 1979 report.[34] Child and Hotchkis remarked:

> The TOTEM tests were reported to have been designed to test the effects of ^{240}Pu [plutonium-240] on weapon yield and performance. No detail of the plutonium ratios in these weapons nor on dispersal of fallout from these tests exists within the public domain, other than to note that different proportions of ^{240}Pu were present in the two test designs and that the wind at the time of detonation was blowing from 211° to 240° dispersing the fallout cloud

to the Northern coast of Australia near Townsville ~ 50 h[ours] later.[35]

For all the intense secrecy around the tests, leaving evidence of the atomic fuel on the ground for future scientists to find seems contradictory as well as unethical. Later analysis of the residue has allowed scientists to obtain some (incomplete) insights into what was in those bombs. Child and Hotchkis found, upon examining the ratio of plutonium-239 to plutonium-240 at the Totem I ground zero, signs that 'short irradiation times were used in production of the weapon material, governed perhaps by fears that the presence of significant ^{240}Pu [plutonium-240] might cause premature fissioning'.[36] They also found when comparing the ground zeroes of Totem I and Totem II that the two sites were quite different, most likely caused by the different yields of the two explosions.[37]

The 500 square kilometres of the Emu lands and a 200-square-kilometre strip to the west were returned to the Traditional Owners by the South Australian Government in 1998, well before the Maralinga test site was returned to its owners.[38] The site is maintained by Maralinga Tjarutja, and no-one can enter it without permission from both Maralinga Tjarutja and the Department of Defence.

The question of the extent to which the atomic trials at Emu were connected to the British rocket tests at Woomera, while interesting, is ultimately a dead-end. The organisations that ran the two operations were separate, although they were both within the UK Ministry of Supply. Certainly early negotiations for the creation of the Joint Project at the Woomera rocket range, going back as far as 1946, seemed to hint that future British rocket technology would eventually be joined with atomic weapons technology.[39] The fact that the atomic tests were also held on the Woomera range suggests that it is likely conversations were had about how the devices tested in Totem could be joined to the long range rockets the British were developing, although in the end that did not occur. The work of both projects had obvious synergies, and Britain clearly wanted nuclear-armed ballistic missiles. Soon, though, American technology would overtake these early plans. Around the time of Totem, the prospect of British-designed, atomic armed ballistic missiles was openly discussed. The Adelaide *Mail* ran a story from London in December 1952 that speculated about where the Woomera trials seemed to be heading:

> Another nuclear field in which British physicists are confident, the new year will show a significant advance in the development of guided missiles carrying atomic warheads ... The new weapons will almost certainly go to Australia for further testing in the coming year.[40]

Just after the first Totem test, the *West Australian* followed a similar line:

> These are secret matters on which it is fruitless for laymen to speculate and they will have to be content with such official information as may be divulged. One guess that may perhaps be hazarded, however, is that trials are being made to perfect a small atomic weapon which can be fitted as a warhead of the guided missiles in which there is some reason to believe that an Anglo–Australian lead is being established over the rest of the world.[41]

Although some work was done at Woomera on the casings of both atomic and thermonuclear weapons, it went no further than mocking up the bomb casing designs in magnesium.[42] Blue Danube was never attached to a British missile. It was adapted for use in V bombers and deployed to RAF facilities, until it was removed from service in 1962. Both Blue Danube and the more compact Red Beard (the latter developed at Maralinga) were too heavy to be carried by missiles in any case.[43]

In 1959 the British minister for defence, Duncan Sandys, asked Australia to consent to the testing of rockets at Woomera that carried nuclear warheads – not complete, armed warheads (though they would carry a high explosive charge). Only the mechanical and electrical components of nuclear warheads were attached to the experimental missiles. By this time Anglo–US fences had been mended, and the design was to be American, not British. Crucially, 'access to the United States atomic information incorporated in the warheads will be restricted to United Kingdom personnel only, special arrangements being made at W.R.E. [Weapons Research Establishment, Salisbury SA] as necessary to secure this'.[44]

The Australians had long wanted their two major Western allies to patch things up and work together again on nuclear armaments, so they couldn't regret it when it happened. Still, it did intensify their outsider status when Britain proposed to carry out further research and development in Australia without giving Australia any rights at all. Defence Minister Townley and Deputy Prime Minister John McEwen outlined their reservations in a top-secret memo to Prime Minister Robert Menzies in 1959:

> First of all we have always criticised the lack of interdependence in this field amongst the western allies and really applauded the U.S. amendment to the McMahon Act which gave the United Kingdom access to United States

research on nuclear weapons. Clearly therefore we cannot bluntly reject the United Kingdom proposal out of hand even if we were so disposed. Furthermore the Department of Supply sees no substantial difficulty in conducting trials under arrangements which would restrict information on nuclear warheads to United Kingdom personnel. However, this is subject to reservation that safety must remain Australian responsibility and that all information relevant to the determination of the necessary safety measures must be made available to Australia.[45]

They also considered the value of the exchange:

Australia has made considerable contributions to nuclear and guided weapons without a great deal of specific benefit to herself. When uranium ore was a limiting factor Australia sold ore without strings; the geographical and political environment for nuclear trials has been provided without any direct benefits, other than access to 'weapon effects' information, to our own services, and the programme of the Joint Project has been determined primarily by consideration of the defence of the U.K. It would be seen therefore that we have a good case to maintain what rights we have. Clearly it would warrant some review of the expenditure of £9.5M annually if the 'joint project' were one from which one of the partners was receiving virtually no benefit.[46]

By the late 1950s, the Australian Government was beginning to feel a little used. In his Cabinet submission, Townley was clear that Australia should be a full and trusted partner in the Joint Project, despite the country's experiences to date:

> Even though it is not proposed to equip our Services with any types of nuclear weapons at this stage, it is considered that Australia should retain as far as possible its rights to information under the Joint Project and its freedom of action in the future. Acceptance of Mr. Sandys' letter as it stands [for simulated nuclear warheads of US design to be tested at Woomera] would destroy to a considerable extent the basis of full partnership in the Joint Project and substantially reduce its potential value to Australian defence.[47]

Australia, in effect, would be a convenient location for both America and Britain to do tests well away from their own populations. Although Townley argued strongly that Australia should not depart from or weaken the terms of the Joint Project, on 19 November 1959 Cabinet thought otherwise:

> Cabinet decided against any further action to press for access to information on the nuclear mechanisms which are the subject of the present United Kingdom request. It made this decision largely against the background that to press further would be to run counter to the general policy that it is desirable to limit nuclear weapon capacities to the few major powers.[48]

Australia never really stopped being a lapdog for British weapons development on its territory.

The lack of effective Australian control over the atomic tests was at its most acute during operations Hurricane and Totem. During both these test series, there was no systematic monitoring of fallout at all, a reckless fallout experiment with wide-reaching and still unfathomable consequences.[49] The Royal Commission found that 'people exposed to radiation as a result of their participation in the Totem tests have an increased risk of cancers as a result of that exposure, although the Royal Commission has been unable to quantify the probable increase'.[50]

There were other consequences too. Many veterans of the atomic tests at Monte Bello, Emu Field and Maralinga are, to this day, terrified of the Official Secrets Act, the British law that provides harsh penalties for passing on state secrets. The Australian equivalent was the Crimes Act, but both Australian and British veterans only remember the Official Secrets Act, and its associated jail sentences that shocked them into silence. Men in their 80s and 90s are still frightened by the threats they received when they went to Emu or the other Australian test sites, and many are reluctant to say much about their experiences.

Emu Field was left contaminated when the British departed. It has stayed that way, more or less, ever since. Operation Brumby, conducted by an AWRE team led by scientist Noah Pearce, undertook a cursory and inadequate clean-up of the Emu and Maralinga sites. A subsequent report, which came to be known as the Pearce report and provided to the Australian government in 1968 as an accurate summation of contamination at the sites, outlined the activities of Operation Brumby. The Pearce report was notoriously wrong about the extent of radioactive contamination at all British atomic test sites.[51] However, beginning in the late 1970s, through the Royal Commission of the mid-1980s to later analyses in the early 1990s, Operation Brumby and the Pearce report were found to be deeply flawed. The Pearce report gave only cursory information about the contamination at Emu and, more generally, kept the Australian Government ignorant for a generation.[52] Pearce's major purpose during Operation

Brumby seems to have been a vain attempt to make the test sites – at both Emu and Maralinga – look like they had never been used for atomic testing. At Emu, the major activity, other than removing all fences and signs, was to gather up various kinds of debris and trinitite: 'The aim was to remove the bulk of the glazing and bury it and disperse the remainder'.[53] They hand-scavenged the sites and graded and disc-ploughed (churning up the topsoil using a tractor towing large circular blades)[54] an area of approximately 130 metres radius. They took a similar approach at Maralinga.[55] The ultimate aim was to give all sites the 'appearance of featureless areas of cultivated soil'.[56] The radioactivity they left on the ground was not obvious to the casual visitor.

In May 1991 both Maralinga and Emu Field were listed 'on the Register of the National Estate as being of historical and social significance'.[57] The listing finally placed legal obligations on Australia to protect the sites and prevent habitation, nearly 40 years after Operation Totem. The Emu listing included all the main areas: the claypan and airstrip, the village site at the junction of the Maralinga and Mabel Creek tracks, four Kittens sites 7 kilometres east of the junction, camera tower sites 11.5 and 21 kilometres east-south-east and south-east of the junction respectively, Totem I and Totem II ground zeroes, the track network connecting all the sites and the area generally north-east of the Totem sites containing the downwind range and track network.[58]

In 1993 the British government, after considerable pressure over years from Australia including an angry and revelatory Royal Commission, finally agreed to pay part of the cost of cleaning up the atomic mess they left. After striking this agreement, the Australian Government established the Maralinga Rehabilitation Technical Advisory Committee

(MARTAC) to oversee it.[59] The MARTAC clean-up involved consultation with Aboriginal people of the area and their representatives and, where possible, employment of Aboriginal people. In January 1998 Aboriginal workers erected a line of boundary markers 10 kilometres long at Emu.[60] A plan to scrape contaminated soil into a mound at Emu did not meet with the approval of the Aboriginal community, as it would place an unsightly blot on the landscape, so it was abandoned.[61] There are smaller mounds here and there, but the site was largely bypassed by the Maralinga clean-up in the late 1990s.

In fact, the MARTAC clean-up dealt with Emu Field in a fairly cursory way compared with the much more extensive clean-up program at Maralinga, failing to spend the $600 000 specifically budgeted for it.[62] With assistance from the Australian Nuclear Science and Technology Organisation, MARTAC calculated that although the trinitite glass fragments and beads were radioactive, someone would have to spend about 50 hours holding a fragment near their nose, snipping at it with pliers and all the while inhaling the fine particles released, for any harm to occur. The droll tone of some parts of the MARTAC report were brought to bear on this idea: 'Credible reasons for spending even a moderate amount of time in random breaking of glazing with pliers are difficult to imagine'.[63] Incredibly, MARTAC considered finding a willing entrepreneur who could turn the trinitite into a commercial venture. Since the glassy material only had low rates of radioactivity, 'bulk handling does not lead to a significant dose', and the material had considerable curiosity value as an historical and scientific relic.[64] Such a business venture never did get off the ground. MARTAC's main concern was to ensure that casual visitors would not be inclined to souvenir the stuff (especially if they were keen on using their pliers on it).

Since being handed over to Maralinga Tjarutja in 1996, the Emu Field area has been managed by its council and is patrolled by the Oak Valley Rangers.[65] These days there are mounds of dirt and debris in many places, and the occasional attempt is made to remove the remaining broken and rusted items. The claypan is as dramatic as ever and still vividly orange, even though it is blighted by old bits of rusting metal and other scrap left to moulder on its edges. Every so often, an aeroplane lands there.

The rangers travel out from their base at Oak Valley on lengthy patrols. In all, they cover some 10.6 million hectares in the central far west of South Australia, 11 per cent of the state.[66] This territory is to the south of the Anangu Pitjantjatjara Yankunytjatjara lands and east of Spinifex land. The rangers have 12 major items on their work list: mapping, managing and monitoring weeds; tracking and culling feral camels; undertaking traditional burning activities (*waru*); managing the flora of the area; managing feral cats and foxes; managing visitors including permit checking; looking after facilities and equipment; undertaking various kinds of programs such as Geographic Information Systems mapping and project planning; undertaking cultural site management; managing work health and safety; and providing community support.[67] They have plenty of work to do, and get to Emu a couple of times per year, depending on their commitments. However, neither they nor anyone else stays at this old military outpost for long.

Emu Field is a tiny slice of a vast swathe of beautiful and wild central Australia territory, a place that is intimidating in some ways owing to its hidden water and its unforgiving climate. Its traditional custodians travelled across this land for millennia, inveterate travellers that they were, always on the move. At the time of the tests no-one lived there permanently, and no-one can live there now. The fact that this pristine land was so casually handed over to a foreign power to test weapons of war speaks volumes about the peculiar bonds of colonialism.

Many Aboriginal people lived or travelled within range of the Operation Totem toxins that were released without adequate warning or safety advice. The remnants of this remote and abandoned atomic test station are still there, sitting silently under the blue dome of the sky, although they are not always obvious. The forgetting of Emu began as soon as the last serviceman and peace officer packed up their rough camps and headed home, away from the flies and the heat and dust. The least visible but most dangerous residues are the wisps of long-ago unleashed radioactivity and the immense human suffering and anxiety associated with them. This is Emu Field's secret, sorry and enduring legacy.

Appendix
BRITISH ATOMIC TESTS IN AUSTRALIA

Major trials

Operation Hurricane
Monte Bello Islands, Western Australia
3 October 1952, 25 kilotons yield

Operation Totem
Emu Field, South Australia
Totem 1: 15 October 1953, 9.1 kilotons yield
Totem 2: 27 October 1953, 7.1 kilotons yield

Operation Mosaic
Monte Bello Islands, Western Australia
Mosaic G1: 16 May 1956, 15–20 kilotons yield
Mosaic G2: 19 June 1956, 60–98 kilotons (disputed) yield

Operation Buffalo
Maralinga, South Australia
Buffalo 1 (One Tree): 27 September 1956, 15 kilotons yield
Buffalo 2 (Marcoo): 4 October 1956, 1.5 kilotons yield
Buffalo 3 (Kite): 11 October 1956, 3 kilotons yield
Buffalo 4 (Breakaway): 22 October 1956, 10 kilotons yield

Operation Antler
Maralinga, South Australia
Antler 1 (Tadje): 14 September 1957, 1 kiloton yield
Antler 2 (Biak): 25 September 1957, 6 kilotons yield
Antler 3 (Taranaki): 9 October 1957, 25 kilotons yield

Minor trials
Kittens
Emu Field, South Australia
September–October 1953
Maralinga, South Australia
(Naya): May–June 1955
(Naya): March 1956
(Naya): March–July 1957
(Naya): March–July 1959
(Naya): May 1961

Tims
Maralinga, South Australia
(Naya): July 1955
(Kuli/Naya): March–July 1957
(Kuli): September–November 1957
(Kuli): April–June 1958
(Kuli): September–November 1958
(Kuli): May–November 1959
(Kuli): April–October 1960
(Kuli/Naya): August 1961
(Kuli): March–April 1963

Rats
Maralinga, South Australia
(Naya): April–June 1958
(Naya): September–November 1958
(Dobo): March–July 1959
(Naya and Dobo): September 1960

Vixen A
Maralinga (Wewak), South Australia
June–August 1959
May–August 1960
March–April 1961

Vixen B
Maralinga (Taranaki), South Australia
September–October 1960
April–May 1961
March–April 1963

SOURCE Adapted from the Royal Commission into British Nuclear Tests in Australia, *Report* (JR McClelland, president), vol. 1, AGPS, Canberra, 1985, appendix G, chronology, pp. vii–1 to vii–10.

GLOSSARY

Alpha particles
Positively charged particles containing two protons and two neutrons that are emitted by certain radioisotopes, particularly those with a high atomic number.

Atom
The smallest particle of an element that retains the characteristics of that element. It is made up of a nucleus and a cloud of surrounding electrons.

Black mist
A mysterious phenomenon associated with Totem I at Emu Field, a 9.1-kiloton weapon exploded on 15 October 1953. The black mist was a rolling, moist fog that covered communities in the vicinity of Wallatinna in South Australia approximately five hours after detonation of the weapon.

D-notice
A secret government request to senior media representatives not to publish certain specified details about defence- or security-related activities. The D-notice system was adopted in Australia in 1952. D-notices were decided by the Defence, Press and Broadcasting Committee administered by the Department of Defence and made up of senior government and media representatives.

Dose
The amount of energy delivered to a mass of material by ionising radiation passing through it.

Dosimeter
A device, instrument or system used to measure or evaluate a dose of radiation.

Fallout
The descent to the Earth's surface of particles contaminated with radioactivity, following the dispersion of radioactive material into the atmosphere by nuclear explosion. The term is applied both to the process and, in a collective sense, to the particulate matter.

Film badge
A plastic holder containing a piece of film similar to a dental x-ray film and worn by personnel at a nuclear test. Radiation passes through the paper and exposes the film. After a nuclear test, the film is developed and the degree of darkening apparent is a measure of the radiation dose received. The film holder usually contains metal filters to enable discrimination between different types of radiation.

Fission
The process in which the nucleus of a heavy element such as uranium or plutonium splits into two nuclei of lighter elements, accompanied by the release of substantial amounts of energy.

Fusion
The process in which the nuclei of light elements such as the hydrogen isotopes deuterium or tritium combine to form the nucleus of a heavier element, accompanied by the release of substantial amounts of energy.

Gamma radiation
Penetrating electromagnetic radiation emitted from the nucleus of radioactive elements. This form of radiation is most readily measured by monitoring equipment such as film badges and dosimeters.

Ionising radiation
Radiation that integrates with matter to add or remove electrons from the atoms of the material absorbing it, producing electrically charged (positive or negative) atoms called ions.

Isotopes
Forms of the same element whose nuclei contain different numbers of neutrons and therefore have different mass numbers. Isotopes of an element have nearly identical chemical properties but differ in their nuclear properties. For instance, some isotopes of an element may be radioactive, and others not.

Major trials
Atomic tests conducted at Monte Bello Islands, Emu Field and Maralinga in Australia that involved detonating a complete atomic bomb, resulting in a 'mushroom cloud'.

Minor trials
Hundreds of tests conducted at Emu Field and Maralinga in Australia that involved examining how radioactive materials and atomic weaponry would behave under various conditions such as fire or conventional explosion.

Neutron
A nuclear particle with no electric charge (neutral) and a mass approximately equal to or slightly greater than that of a proton. Neutrons are present in all atoms except those of the lightest isotope of hydrogen.

Nuclide
Species of atoms having a specified number of protons and neutrons in their nuclei. Radionuclides are the radioactive forms of nuclides. They are often expressed as, for example, ^{239}Pu, which shows in numerical form the number of neutrons and hence the form of isotope.

Plutonium (Pu)
A dense, silvery radioactive element that does not occur naturally but is made in a reactor by bombarding uranium with neutrons. It was first produced in 1940. Plutonium has 13 known isotopes. A fissile material, ^{239}Pu can be used as the core of a nuclear weapon.

Radioactivity
The property of certain radionuclides of spontaneously emitting particles and/or x-ray or gamma ray radiation, or of undergoing spontaneous fission. The rate of decay is specific to a given species of radionuclide and cannot be changed by known physical or chemical processes.

Radionuclide
A radioactive nuclide. See Nuclide.

Thermonuclear weapon
A nuclear device that relies on raising the temperature of a mixture of deuterium and tritium nuclei to above 10 million degrees Celsius, at which point nuclear fusion reactions occur. This type of weapon is also known as a hydrogen bomb.

Warhead
The explosive head of a bomb.

Yield
The amount of energy generated by a nuclear explosion, usually expressed in kilotons (for fission devices) or megatons (for fusion devices). A kiloton is equivalent to 1,000 tons of TNT and a megaton is equivalent to one million tons of TNT.

<small>Adapted from a variety of sources, in particular the Royal Commission into British Nuclear Tests in Australia, *Report* (JR McClelland, president), vol. 2, AGPS, Canberra, 1985, pp. II–1 to II–9; and JL Symonds, *A History of British Atomic Tests in Australia*, pp. 585–593.</small>

NOTES

Prologue

Epigraph: Yami Lester, *History and the Land: A Story Told by Yami Lester*, ed. Petronella Water, rev. edn, Institute for Aboriginal Development, Alice Springs, 1990, p. 10.

1. WB MacDougall, 'Detailed Survey of the Jangkuntjara Tribe – Their Traditional Tribal Country and Ceremonial Grounds', sent to superintendent, Woomera, 2 March 1953, Malone files.
2. MacDougall, 'Detailed Survey'.
3. MacDougall, 'Detailed Survey'.
4. Scott Cane, 'Something for the Future: Community Development at Oak Valley, S.A.', Report to Oak Valley Community and Aboriginal and Torres Strait Islander Commission (ATSIC), Ceduna, January 1992, p. 1.
5. Lester, *History and the Land*, p. 22.
6. Lester, *History and the Land*, p. 2.
7. Eve Vincent, 'Knowing the Country', *Cultural Studies Review*, vol. 13, no. 2, September 2007, p. 162.
8. Yalata and Oak Valley Communities with Christobel Mattingley, *Maralinga: The Anangu Story*, Allen & Unwin, Sydney, 2009, p. 1.
9. Heather Goodall, '"The Whole Truth and Nothing But ..."': Some Intersections of Western Law, Aboriginal History and Community Memory', *Journal of Australian Studies*, no. 32, March 1992, p. 112.
10. Goodall, '"The Whole Truth and Nothing But ..."', p. 112.
11. Sitarani Kerin, '"Doctor Do-Good"?: Charles Duguid and Aboriginal Politics, 1930s–1970s', unpublished PhD thesis, ANU, December 2004, p. 60.
12. Heather Goodall, interview with the author, 10 June 2021.
13. Kingsley Palmer, 'Dealing with the Legacy of the Past: Aborigines and Atomic Testing in South Australia', *Aboriginal History*, vol. 14, no. 1/2, 1990, p. 197.
14. Palmer, 'Dealing with the Legacy of the Past', p. 197.
15. Charles Duguid, 'The Central Aborigines Reserve', presidential address, Aborigines Advancement League of South Australia Incorporated, 21 October 1957, Reliance Printing Company, Adelaide, 1957.
16. Phyl & Noel Wallace, *Killing Me Softly: The Destruction of a Heritage*, Thomas Nelson, West Melbourne, 1977, p. 69; Yami Lester, Transcript of proceedings, Royal Commission into British Nuclear Tests in Australia, NAA: A6448, 13, p. 7116.
17. JA Cooke, on behalf of acting regional manager, Department of Aboriginal Affairs South Australian Region, 'Aboriginals and Atomic Tests', Letter to secretary, Department of Aboriginal Affairs, Canberra, 4 September 1980, Malone files.
18. Anne O'Brien, 'Hunger and the Humanitarian Frontier', *Aboriginal History*,

vol. 39, 2015, <press-files.anu.edu.au/downloads/press/p332783/html/article05.xhtml?referer=&page=11>, accessed 28 December 2021.
19   'Australia's Part in Preparations for Atomic Test', Press statement, 25 June 1953, NAA: A6456, R095/012.
20   UK Ministry of Defence, Secret meeting notes, Totem planning, 19 December 1952, DEFE 16/454, NAUK.
21   JA Carruthers, RJ Wilson, PC East & EJ Ellis, AWRE report no. T 10/55, Thermal measurements at Operation Totem, ES 5/125, NAUK.
22   'Unofficial Background Scientific Information for Press Observers', Briefing notes supplied to Ernest Titterton, [1953], Titterton archive, box 100, series 22, AAS.
23   Ivan Southwell, *Woomera*, Angus & Robertson, Sydney, 1962, p. 160.
24   Southwell, *Woomera*, p. 160.
25   Peter Walsh, *Confessions of a Failed Finance Minister*, Random House Australia, Milsons Point, 1995, p. 96.
26   Indigenous people from Wallatinna, Transcript of proceedings, Royal Commission into British Nuclear Tests in Australia, NAA: A6448, 13, pp. 7081–202.
27   Garry Hiskey, *Maralinga: The Struggle for the Return of the Lands*, Wakefield Press, Adelaide, 2021, p. 1.
28   Richard Bradshaw, former Pitjantjatjara Council lawyer, interview with the author, 25 May 2021.
29   William Penney, Transcript of proceedings, Royal Commission into British Nuclear Tests in Australia, NAA: A6448, 9, p. 4345.

Chapter 1 Finding X200

Epigraph: 'A British Nevada', *Advertiser* (Adelaide), 26 June 1953, p. 2.
1   DS Maclagen, MB Cooper & JC Duggleby, 'Residual Radioactive Contamination of Emu Plain from Nuclear Weapons Tests Conducted in 1953', Draft ARL report, August 1979, NAA: A6465, R057/003.
2   Royal Commission into British Nuclear Tests in Australia, *Report* (JR McClelland, president), 2 vols, AGPS, Canberra, 1985, p. 463.
3   ER Drake Seager, Army Equipment Group, War Office, Report on Operation Totem, 14 December 1953, ES 1/834, NAUK.
4   Drake Seager, Report.
5   Ministry of Supply meeting, Top-secret note, Ref. 330/10/88/1, St Giles Court, London, 28 February 1953, AB 16/1743, NAUK.
6   JL Symonds, *A History of British Atomic Tests in Australia*, AGPS, Canberra, 1985, p. 115.
7   Totem Panel, Secret note on Ben Cockram's letter to Allen Brown of 21 January 1953, 6 February 1953, NAA: A6456, R124/018.
8   L Martin, A Butement & W Penney, Notes on an 'off the record' talk between Professor L Martin, Dr Butement and Dr WG Penney, 24 April 1952, NAA: A5954, 7/11.
9   Martin, Butement & Penney, Notes.
10 Martin, Butement & Penney, Notes.
11 Royal Commission, *Report*, p. 137.
12 'Britain "A Factor in A-Warfare"', *Argus*, 9 January 1953, Department of Defence, Atomic energy – general, NAA: A5954, 2163/1.
13 'Britain "A Factor in A-Warfare"'.

14 Royal Commission, *Report*, p. 137.
15 UK *aide memoire* to Australia, 13 December 1952, reproduced in Symonds, *A History of British Atomic Tests*, p. 126.
16 Australian reply to UK *aide memoire*, 15 December 1952, reproduced in Symonds, *A History of British Atomic Tests*, p. 127.
17 Eve Vincent, 'Knowing the Country', *Cultural Studies Review*, vol. 13, no. 2, September 2007, p. 161.
18 William Penney, Secret letter to JES Stevens, chairman, Totem Panel, 2 November 1953, NAA: A6456, R124/018.
19 Ivan Southwell, *Woomera*, Angus & Robertson, Sydney, 1962, p. 170.
20 Len Beadell, Transcript of proceedings, Royal Commission into British Nuclear Tests in Australia, NAA: A6448, 7, p. 3501.
21 Beadell, Transcript of proceedings, p. 3501.
22 Ian Bayly, *Len Beadell's Legacy: Australia's Atomic Bomb and Rocket Roads*, Bas Publishing, Seaford, 2010, p. 21.
23 Bayly, *Len Beadell's Legacy*, pp. 21–23.
24 Beadell, Transcript of proceedings, p. 3502.
25 Beadell, Transcript of proceedings, p. 3502.
26 Bayly, *Len Beadell's Legacy*, p. 163.
27 Department of Education, Science and Training, *Rehabilitation of Former Nuclear Tests Sites at Emu and Maralinga (Australia) 2003* (MARTAC report), Commonwealth of Australia, Canberra, 2002, p. 6.
28 Ben Cockram, Top-secret letter to Neil Pritchard, 19 January 1953, AB 16/1743, NAUK.
29 Cockram, Top-secret letter to Pritchard.
30 Cockram, Top-secret letter to Pritchard.
31 AS Brown, secretary, Prime Minister's Department, Top-secret personal note to Ben Cockram, UK High Commission, 18 March 1953, NAA: A6456, R124/018.
32 Prime Minister's Department, Top-secret note, 13 March 1953, NAA: A6456, R124/018.
33 Prime Minister's Department, Top-secret note.
34 Prime Minister's Department, Top-secret note.
35 Ben Cockram, Top-secret letter to Robert Menzies, 3 March 1953, NAA: A6456, R124/018.
36 AW Fadden, acting prime minister, Draft memorandum to all ministers, 22 June 1953, NAA: A6456, R124/018.
37 CA Stinson, acting secretary, Department of Supply, Confidential memorandum to director, Commonwealth Investigation Service, Attorney-General's Department, 14 January 1954, NAA: A1533, 1954/299.
38 Stinson, Confidential memorandum.
39 James McClelland, *Stirring the Possum: A Political Autobiography*, Penguin, Ringwood, 1988, p. 213.
40 Ernest Titterton, Secret letter to Allen Brown, secretary, Prime Minister's Department, [early 1953], NAA: A6456, R215/016.
41 Titterton, Secret letter.
42 Titterton, Secret letter.
43 Titterton, Secret letter.
44 Titterton, Secret letter.

NOTES TO PAGES 32-40

45 William Penney, Personal note to Ernest Titterton, 22 May 1953, Titterton archive, box 100, series 22, AAS.
46 AD McKnight, Prime Minister's Department, File note, 29 April 1953, NAA: A6456, R124/018.
47 McKnight, File note.
48 William Penney, Top-secret letter to Lord Cherwell, undated, NAA: A6455, AB8.
49 Howard Beale, minister for supply, 'Atomic Tests in Australia', Top-secret Cabinet briefing document, Submission no. 73, Canberra, 11 August 1954, Malone files.
50 McKnight, File note.
51 JF Hill, Report of enquiry into souveniring of official stores from Emu Field, addressed to deputy director, Melbourne, Commonwealth Investigation Service, 25 April 1954, NAA: A1533, 1954/299.
52 Hill, Report.
53 Hill, Report.
54 LG Egan, assistant crown solicitor, 'Woomera Rocket Range', Memorandum to assistant secretary, R&D, Department of Supply, 15 April 1953, NAA: A6456, R215/016.
55 Egan, 'Woomera Rocket Range'.
56 Leo Carter, Top-secret memorandum to chairman, Totem Panel, 31 March 1953, NAA: A6456, R215/016.
57 Carter, Top-secret memorandum.
58 William Penney, Restricted de-briefing notice for UK staff, undated, Titterton archive, box 100, series 22, AAS.
59 Penney, Restricted de-briefing notice.
60 Hill, Report.
61 E Hattam, deputy superintending peace officer, Commonwealth Investigation Service, Canberra, Letter to superintending peace officer, 28 May 1953, NAA: A6456, R121/100.
62 Hattam, Letter.
63 Hattam, Letter.
64 Duties of Mabel Creek Detachment, Statement of duties, 11 June 1953, NAA: A6456, R121/100.
65 Duties of Mabel Creek Detachment.
66 Duties of Mabel Creek Detachment.
67 CS Morrison, X200 project security officer, Post orders – Peace Officer Guard – X200 site Emu, NAA: A6456, R121/100.
68 CA Adams, Letter to Ernest Titterton, 11 September 1953, Titterton archive, box 100, series 22, AAS.
69 TWJ Redmond, UK staff, Letter to Ernest Titterton, 11 September 1953, Titterton archive, box 100, series 22, AAS.
70 Defence Committee, Atomic test – Montebello, Notes of meeting of Dr Solandt, chairman, Canadian Defence Research Board, & Professor Martin, defence scientific adviser, with Defence Committee, 9 October 1952, NAA: A5954, 7/11.
71 Defence Committee, Atomic test.
72 William Penney, Transcript of proceedings, Royal Commission into British Nuclear Tests in Australia, NAA: A6448, 9, p. 4336.

73   Peter Morton, *Fire across the Desert: Woomera and the Anglo-Australian Joint Project 1946–1980*, Defence Science and Technology, Department of Defence, Canberra, 1989, p. 43.

Chapter 2 Why Emu?

Epigraphs: Royal Commission into British Nuclear Tests in Australia, *Report: Conclusions and Recommendations* (JR McClelland, president, J Fitch and WJA Jonas, commissioners), AGPS, Canberra, 1985, p. 7; Paul Malone & Howard Conkey, 'Impact on Aborigines: The Loud Bang and the Black Cloud Linger', *Canberra Times*, 29 September 1984, p. 15.

1   Captain FB Lloyd, Ministry of Supply, London, Top-secret letter to Neil Pritchard, Commonwealth Relations Office, 7 December 1953, FCO 1/6, NAUK.
2   Lloyd, Top-secret letter.
3   Neil Pritchard, Top-secret letter to Ben Cockram, 25 November 1953, FCO 1/6, NAUK.
4   William Penney, Transcript of proceedings, Royal Commission into British Nuclear Tests in Australia, NAA: A6448, 9, p. 4337.
5   Penney, Transcript of proceedings, p. 4338.
6   High Explosives Research, Notes of meeting, 28 November 1952, DEFE 16/454, NAUK.
7   High Explosives Research, Notes.
8   High Explosives Research, Notes.
9   CN Hill, *An Atomic Empire: A Technical History of the Rise and Fall of the British Atomic Energy Programme*, Imperial College Press, London, 2013, p. 160.
10 Margaret Gowing with Lorna Arnold, *Independence and Deterrence: Britain and Atomic Energy, 1945–1952*, UK Atomic Energy Authority, St Martin's Press, New York, 1974, p. 321.
11 Lorna Arnold & Mark Smith, *Britain, Australia and the Bomb: The Nuclear Tests and Their Aftermath*, 2nd edn, Palgrave Macmillan, New York, 2006, p. 49.
12 Hill, *An Atomic Empire*, p. 162.
13 Hill, *An Atomic Empire*, p. 161.
14 Official Committee on Atomic Energy, CAB 134/748 PRO, 20 January 1953, quoted in Arnold & Smith, *Britain, Australia and the Bomb*, p. 50.
15 William Penney, Top-secret letter to FC How, UK Ministry of Supply, 18 December 1952, AB 16/1743, NAUK.
16 Penney, Top-secret letter.
17 Penney, Top-secret letter.
18 Penney, Top-secret letter.
19 Department of Supply, Souveniring of official stores at Emu Field, 18 December 1953, NAA: A6456, R086/011.
20 LC Tyte, AWRE, Meeting with visiting Australian mission, February 1953, Minutes dated 14 April 1953, AB 16/4544, NAUK.
21 Tyte, Meeting.
22 Tyte, Meeting.
23 Tyte, Meeting.
24 Tyte, Meeting.
25 Tyte, Meeting.

26 Tyte, Meeting.
27 Tyte, Meeting.
28 Frederick Lindemann (Lord Cherwell), Note to Thomas Elmhirst, 19 January 1953, ELMT 4/2, Elmhirst files, Churchill Archives Centre, University of Cambridge.
29 JM Wilson, assistant secretary, Establishments Division, UK Ministry of Supply, Personal and confidential letter to Thomas Elmhirst, 28 January 1953, ELMT 4/2, Elmhirst files, Churchill Archives Centre, University of Cambridge.
30 Thomas Elmhirst, Confidential note to JM Wilson, assistant secretary, Establishments Division, 7 October 1953, ELMT 4/2, Elmhirst files, Churchill Archives Centre, University of Cambridge.
31 Elmhirst, Confidential note.
32 Earl Marshal's Office, Belgrave Square, London, Letter to Elmhirst, 16 January 1953, ELMT 4/2, Elmhirst files, Churchill Archives Centre, University of Cambridge.
33 Ministry of Supply meeting, Top-secret note, Ref. 330/10/88/1, St Giles Court, London, 28 February 1953, AB 16/1743, NAUK.
34 Press handout material, Totem Panel, Minutes, Eighth meeting, Swanston Street, Melbourne, 24 July 1953, NAA: A6456, R095/012.
35 Totem Panel, Minutes, Sixth meeting, Swanston Street, Melbourne, 26 May 1953, NAA: A6456, R096/002.
36 Totem Panel, Minutes, Eighth meeting.
37 Totem Panel, Minutes, Eighth meeting.
38 Editorial, *Bulldust* (Emu Field), 8 August 1953, collection of Wesley Perriman, curator, British Nuclear Tests Veterans' Association.
39 Totem Panel, Minutes, Eighth meeting.
40 Totem Panel, Minutes, Eighth meeting.
41 Lieutenant-General Sir Alwyn Garrett, chief of the general staff, Letter to Major General Eric Woodward, adjutant-general, Australian Army, 24 June 1953, NAA: A6456, R030/174.
42 Garrett, Letter.
43 'Welcome to the Claypan', An introduction to living and working at the Emu claypan campsite, 1953, NAA: D2861, WELCOME TO THE CLAYPAN.
44 'Atomic Test Base Workers Held Coronation Parade', Press statement, undated, NAA: A6456, R095/012.
45 Project X200 progress report no. 11 for week ended 6 June 1953, Top-secret note to chairman, Totem Panel, 10 June 1953, NAA: A6456, R096/002.
46 Project X200 progress report.
47 LC Lucas, director of construction, X200, Secret letter to secretary, Department of Supply, Melbourne, 5 March 1953, DEFE 16/679, NAUK.
48 John Simpson, The Independent Nuclear State: The United States, Britain and the Military Atom, Palgrave Macmillan, London, 1983, p. 91.
49 JL Symonds, *A History of British Atomic Tests in Australia*, AGPS, Canberra, 1985, p. 117.
50 Symonds, *A History of British Atomic Tests*, p. 117.
51 UK high commissioner, Telegram to Commonwealth Relations Office, 28 February 1953, AB 16/1743, NAUK.
52 UK high commissioner, Telegram.
53 Neil Pritchard, British deputy high commissioner to Australia, Top-secret letter to Ben Cockram, 2 March 1953, AB 16/1743, NAUK.

54 Pritchard, Top-secret letter.
55 Pritchard, Top-secret letter.
56 Secretary, Aborigines Protection Board, Adelaide, Memorandum to Walter MacDougall, NPO, 26 August 1953, Malone files.
57 'Australia's Part in Preparations for Atomic Test', Press statement, 25 June 1953, NAA: A6456, R095/012.
58 'Man Who Led Fighting Sixth Is Dead', *Age*, 21 May 1969, Personal correspondence – Australia, Sir JES Stevens, secretary, Department of Supply, NAA: A5954, 70/11.
59 Atomic warfare resources, Report no. 32/1949, Defence Research and Development Policy Committee, 7 December 1949, NAA: A5954, 2330/2.
60 Atomic warfare resources.
61 Katherine Pyne, 'Art or Article? The Need for and Nature of the British Hydrogen Bomb, 1954–58', *Contemporary British History*, vol. 9, no. 3, 1995, p. 567.
62 Pyne, 'Art or Article?', p. 568.
63 Pyne, 'Art or Article?', p. 568.
64 Atomic warfare resources.

Chapter 3 Towards a British 'Austerity Bomb'
Epigraph: Martin Theaker, 'Being Nuclear on a Budget: Churchill, Britain and "Atoms for Peace" 1953–1955', *Diplomacy & Statecraft*, vol. 27, no. 4, 2016, p. 655.
1 Margaret Gowing with Lorna Arnold, *Independence and Deterrence: Britain and Atomic Energy, 1945–1952*, UK Atomic Energy Authority, St Martin's Press, New York, 1974, p. 407.
2 EM Fitzgerald, 'Allison, Attlee and the Bomb: Views on the 1947 British Decision to Build an Atom Bomb', *RUSI Journal*, vol. 122, no. 1, 1977, p. 49.
3 Lord Portal, Timeline, Lord Portal archive, Atomic energy folder B, UK military atomic energy chronology 1945–1969 (including V bomber force), Portal Collection, Christ Church Library, Oxford.
4 Portal, Timeline.
5 Alfred Goldberg, 'The Atomic Origins of the British Nuclear Deterrent', *International Affairs* (Royal Institute of International Affairs), July 1964, p. 418.
6 Goldberg, 'The Atomic Origins', p. 426.
7 R Berman, 'Lindemann in Physics', *Notes and Records of the Royal Society of London*, vol. 41, no. 2, June 1987, p. 181.
8 Madhusree Mukerjee, 'Lord Cherwell: Churchill's Confidence Man', HistoryNet, <www.historynet.com/lord-cherwell-churchills-confidence-man.htm>, accessed 5 March 2020.
9 Madhusree Mukerjee, 'The Most Powerful Scientist Ever: Winston Churchill's Personal Technocrat', Scientific American, 6 August 2010, <www.scientificamerican.com/article/the-most-powerful-scientist-ever/>, accessed 27 August 2021.
10 Mukerjee, 'Lord Cherwell'.
11 Mukerjee, 'Lord Cherwell'.
12 Gowing with Arnold, *Independence and Deterrence*, p. 407.
13 Stephen Twigge, 'The Atomic Marshall Plan: Atoms for Peace, British Diplomacy and Civil Nuclear Power', *Cold War History*, vol. 16, no. 2, 2016, p. 220.

14   Portal, Timeline.
15   Portal, Timeline.
16   Portal, Timeline.
17   Lord Portal, Transcribed notes provided to Margaret Gowing in response to her questions, Lord Portal archive, Atomic energy folder B, UK military atomic energy chronology.
18   Gowing, quoted in Portal, Transcribed notes.
19   James Chadwick, Handwritten letter to Lord Portal, 24 October 1966, in response to a letter for Chadwick's 75th birthday, Lord Portal archive, Atomic energy folder B2, UK military atomic energy chronology.
20   James Chadwick, 'Private and Personal', Top-secret letter to Lord Portal, 8 April 1946, Lord Portal archive, Atomic energy folder B2, UK military atomic energy chronology.
21   James Chadwick, Notes on draft memorandum by Portal, 5 August 1946, Lord Portal archive, Atomic energy folder B, UK military atomic energy chronology.
22   Portal, Transcribed notes.
23   Chadwick, Notes.
24   Chadwick, Notes.
25   Chadwick, Private and Personal letter to Portal.
26   RAF Barnham Nuclear Weapon Storage Site, 'Blue Danube Free Fall Nuclear Bomb', RAF Barnham Nuclear Weapon Storage Site, <rafbarnham-nss.weebly.com/blue-danube.html>, accessed 28 November 2021.
27   Chadwick, Notes.
28   Chadwick, Notes.
29   Chadwick, Notes.
30   Chadwick, Notes.
31   J Wilmot, minister of supply, Memorandum to Lord Portal, Minutes of the GEN.163 meeting, 10 Downing Street, London, 8 January 1947, Lord Portal archive, Atomic energy folder B, UK military atomic energy chronology.
32   Quebec Agreement, reproduced in Margaret Gowing, *Britain and Atomic Energy 1939–1945*, London, Macmillan, 1964, pp. 439–40.
33   George Strauss, Ministry of Supply – Atomic Bomb, from unpublished autobiography, STRS 1/2 Strauss papers, Churchill Archive Centre, University of Cambridge, p. 124.
34   Strauss, Ministry of Supply, pp. 124–25.
35   Strauss, Ministry of Supply, p. 127.
36   Harry S Truman, 'Statement by the President of the United States', 6 August 1945, transcription in 'Announcing the Bombing of Hiroshima', American Experience, <www.pbs.org/wgbh/americanexperience/features/primary-resources/truman-hiroshima>, accessed 27 September 2021.
37   Strauss, Ministry of Supply, p. 127.
38   Kevin Ruane *Churchill and the Bomb in War and Cold War*, Bloomsbury Academic, London, 2016, p. 209.
39   Theaker, 'Being Nuclear on a Budget', p. 639.
40   Draft press notice, 19 February 1951, annotated by Winston Churchill, CHUR 2/28, Churchill Archives Centre, University of Cambridge.
41   Gowing with Arnold, *Independence and Deterrence*, p. 406.
42   Graham Farmelo, *Churchill's Bomb: A Hidden History of Science, War and Politics*, Faber & Faber, London, 2013, p. 381.
43   Farmelo, *Churchill's Bomb*, p. 383.

44 Gowing with Arnold, *Independence and Deterrence*, p. 409.
45 Gowing with Arnold, *Independence and Deterrence*, p. 410.
46 Richard Gott, 'The Evolution of the Independent British Deterrent', *International Affairs* (Royal Institute of International Affairs), vol. 39, no. 2, April 1963, p. 239.
47 Kevin Ruane, *Churchill and the Bomb in War and Cold War*, Bloomsbury Academic, London, 2016, p. 313.
48 Duncan Sandys, *Defence: outline of future policy* (Defence White Paper), Her Majesty's Stationery Office, London, March 1957, NAUK, CAB/129/86.
49 Air Commodore DN Roberts, director of operations, Top-secret note to DCAS, 15 January 1953, AIR 2/13778, NAUK.
50 William Penney, Top-secret letter to Air Chief Marshal Sir Ralph Cochrane, 29 August 1952, AIR 2/13778, NAUK.
51 Penney, Top-secret letter.
52 William Penney, Transcript of interview with Alfred Goldberg, United States Air Force Historical Division, 16 June 1963, AIR 2/13778, NAUK.
53 Penney, Transcript of interview.
54 Air Marshall RO Jones, chair, SALOME Committee, Top-secret minute to ACAS, 22 June 1953, AIR 2/13778, NAUK.
55 HEROD Committee, Minutes of the 8th meeting held at Air Ministry, Whitehall, chaired by Sir Ronald Ivelaw-Chapman, 20 July 1953, AIR 2/13778, NAUK.
56 HEROD Committee, Agenda for eighth meeting to be held on 20 July 1953, AIR 2/13778, NAUK.
57 HEROD Committee, Agenda.
58 HEROD Committee, Agenda.
59 Ruane, *Churchill and the Bomb*, p. 218.
60 Ruane, *Churchill and the Bomb*, p. 219.
61 Ruane, *Churchill and the Bomb*, p. 311.
62 Frederick Lindemann, Secret letter to Winston Churchill, 29 December 1952, PREM 11/290, NAUK.
63 Lindemann, Secret letter.
64 Royal Commission into British Nuclear Tests in Australia, *Report* (JR McClelland, president), 2 vols, AGPS, Canberra, 1985, p. 506.
65 Jonathan Aylen, 'First Waltz: Development and Deployment of Blue Danube, Britain's Post-War Atomic Bomb', *International Journal for the History of Engineering and Technology*, vol. 85, no. 1, January 2015, p. 35.
66 William Penney, Transcript of proceedings, Royal Commission into British Nuclear Tests in Australia, NAA: A6448, 9, p. 4299.
67 Penney, Transcript of proceedings, p. 4299.
68 Penney, Transcript of proceedings, p. 4299.
69 Department of Education, Science and Training, *Rehabilitation of Former Nuclear Tests Sites at Emu and Maralinga (Australia) 2003* (MARTAC report), Commonwealth of Australia, Canberra, 2002, p. 11.
70 LC Tyte, AWRE, Meeting with visiting Australian mission, February 1953, Minutes dated 14 April 1953, AB 16/4544, NAUK.
71 Tyte, Meeting.
72 Tyte, Meeting.
73 Tyte, Meeting.

74 GA Williams, radiation scientist, correspondence with the author, 30 August 2021.
75 UK Ministry of Supply, Operation 'TOTEM': summary plan (prepared by J McEnhill), part 1.
76 McEnhill, Operation Totem summary plan.
77 Carter, Top-secret teleprinter message to Flannery, ROCLABS, from Melbourne, 23 September 1953, DEFE 16/679, NAUK.
78 McEnhill, Operation Totem summary plan.
79 McEnhill, Operation Totem summary plan.
80 Penney, Transcript of interview.
81 Cabinet meeting of ministers, Secret minutes, Cabinet Office, Great George Street, SW1, 12 November 1952, CAB 130/79/GEN417, NAUK.
82 Cabinet meeting, Secret minutes.
83 Cabinet meeting, Secret minutes.
84 Cabinet meeting, Secret minutes.
85 Cabinet meeting, Secret minutes.
86 Gowing with Arnold, *Independence and Deterrence*, p. 408.
87 Cabinet meeting, Secret minutes.
88 Frederick Lindemann, Note to Churchill, 12 November 1953, Lindemann/Cherwell papers, J138, Nuffield College Library, Oxford.
89 Lindemann, Note.

Chapter 4 Sound and fury at Emu
Epigraph: 'A Flash and a Roar', *Advertiser* (Adelaide), 16 October 1953, p. 2.
1 Unseasonal rain: Ray Acaster, 'Sherlock Holmes and the Grim Reaper of Wallatinna: Aboriginal Oral History Solves a Scientific Mystery', *Oral History Association of Australia Journal*, no. 24, 2002, p. 22; delay of test: JA Carruthers, RJ Wilson, PC East & EJ Ellis, AWRE report no. T 10/55, Thermal measurements at Operation Totem, ES 5/125, NAUK.
2 UK Ministry of Defence, Secret memorandum, Totem planning, Outline of site location and layout, DEFE 16/454, NAUK.
3 UK Ministry of Defence, Electronic measurements at Tom-Cat, 18 December 1952, DEFE 16/454, NAUK.
4 UK Ministry of Defence, Secret memorandum, Draft preliminary outline of contribution required by Australian authorities, 12 December 1952, DEFE 16/454, NAUK.
5 High Explosives Research, Notes of meeting, 28 November 1952, DEFE 16/454, NAUK.
6 Frederick Lindemann (Lord Cherwell), Secret note to Mr FC How, 6 February 1953, AB 16/1743, NAUK.
7 FC How, UK Ministry of Supply, Official Committee on Atomic Energy, Top-secret briefing document on Operation Totem, 2 February 1953, AB 16/1743, NAUK.
8 UK Ministry of Defence, Electronic measurements at Tom-Cat.
9 Edward Siddons, Transcript of proceedings, Royal Commission into British Nuclear Tests in Australia, NAA: A6448, 10, p. 5414.
10 William Penney, Transcript of proceedings, Royal Commission into British Nuclear Tests in Australia, NAA: A6448, 9, p. 4341.
11 DS Maclagen, MB Cooper & JC Duggleby, 'Residual Radiation Contamination of Emu Plain from Nuclear Weapons Tests Conducted in 1953', Australian Radiation Laboratory, August 1979, NAA: A6465, R057/003, p. 3.

12 DEHCD, Report.
13 DEHCD, Report.
14 Royal Commission into British Nuclear Tests in Australia, *Report* (JR McClelland, president), 2 vols, AGPS, Canberra, 1985, p. 146.
15 Lorna Arnold & Mark Smith, *Britain, Australia and the Bomb: The Nuclear Tests and Their Aftermath*, 2nd edn, Palgrave Macmillan, New York, 2006, p. 60.
16 Arnold & Smith, *Britain, Australia and the Bomb*, p. 60.
17 Penney, Transcript of proceedings, p. 4354.
18 Department of Education, Science and Training, *Rehabilitation of Former Nuclear Tests Sites at Emu and Maralinga (Australia) 2003* (MARTAC report), Commonwealth of Australia, Canberra, 2002, p. 165.
19 EW Walker, AWRE report no. R67/54, Operation Totem photographic observations, February 1955, ES 5/68, NAUK, p. 10.
20 DEST, *Rehabilitation*, p. 165.
21 Thomas Elmhirst, Top-secret note to FC How, UK Ministry of Supply, 12 March 1953, AB 16/1743, NAUK.
22 Thomas Elmhirst, Note to controller of Atomic Energy, 12 March 1953, AB 16/1743, NAUK.
23 Service department requirements for tests in future atomic bomb trials, Notes from meeting of Defence Policy and Requirements Sub-Committee, (AES)/M(53), 20 January 1953, AB 16/1743, NAUK.
24 Royal Commission, *Report*, p. 226.
25 J McEnhill, UK Ministry of Supply, Operation Totem summary plan, part 1, NAA: A6454, ZB1.
26 Arnold & Smith, *Britain, Australia and the Bomb*, p. 61.
27 John Clarke QC, Air Marshal Doug Riding & Dr David Rosalky, 'British Atomic Tests', in *Report of the Review of Veterans' Entitlements*, vol. 2, Commonwealth of Australia, Canberra, 6 January 2003, p. 391.
28 Lt Col W de LM Messenger, AWRE report no. T78/54, Operation Totem: the effects of an atomic explosion on a Centurion tank, vol. 1, December 1956, ES 5/77, NAUK.
29 McEnhill, Operation Totem summary plan.
30 Royal Commission, *Report*, p. 197.
31 Walker, AWRE report, p. 12.
32 Walker, AWRE report, p. 12.
33 CA Adams, AWRE, Document showing Totem staff, [1953], AB 16/1743, NAUK.
34 FA O'Connor, secretary, Department of Supply, Secret memo to minister of supply, 21 August 1953, DEFE 16/685, NAUK.
35 JES Stevens, Department of Supply, Melbourne, Secret teleprinter messages to Elmhirst, 29 July 1953, DEFE 16/679, NAUK.
36 Ian Bayly, *Len Beadell's Legacy: Australia's Atomic Bomb and Rocket Roads*, Bas Publishing, Seaford, 2010, p. 32.
37 William Cook, Secret cable to Brigadier Lucas, 20 July 1953, DEFE 16/679, NAUK.
38 Service department requirements.
39 Service department requirements.
40 Atomic Weapons Research Establishment, *Operation Totem*, film, Aldermaston, 1953, available at International War Museums, <www.iwm.org.uk/collections/item/object/1060036296>, accessed 2 February 2021.

41 AWRE, *Operation Totem*.
42 GA Williams, Radioactive Waste Safety, 'Maralinga Rehabilitation Project – The Clean-Up at Emu (and What It Means for the Future)', internal, unpublished project summary, Australian Radiation Protection and Nuclear Safety Agency (ARPANSA), 1 August 2017, p. 3.
43 DEST, *Rehabilitation*, p. 165.
44 McEnhill, Operation Totem summary plan.
45 SO book 130, Handwritten diary containing Totem data, ES 18/24, NAUK.
46 SO book 130, Handwritten diary.
47 SO book 130, Handwritten diary.
48 TA Brinkley, 'Atomic Bomb Test Series at X200 Site October 1953: Recollections of Totem I', Titterton archive, box 99, series 22, AAS.
49 Brinkley, 'Atomic Bomb Test Series'.
50 Brinkley, 'Atomic Bomb Test Series'.
51 WR Blunden & WF Caplehorn, Department of Army Report, quoted in JL Symonds, *A History of British Atomic Tests in Australia*, AGPS, Canberra, 1985, p. 184.
52 Blunden & Caplehorn, Report, p. 190.
53 Blunden & Caplehorn, Report, p. 185.
54 E Drake Seager, group leader, AE Group, Report prepared for director, AWRE, 29 January 1954, ES 18/25, NAUK.
55 Drake Seager, Report.
56 Drake Seager, Report.
57 McEnhill, Operation Totem summary plan.
58 Drake Seager, Report.
59 Carruthers, Wilson, East & Ellis, AWRE report.
60 Carruthers, Wilson, East & Ellis, AWRE report.
61 JES Stevens, Top-secret letter to Ernest Titterton, 17 April 1953, NAA: A6456, R215/016.
62 Thomas Elmhirst, head of TOTEX, Top-secret cable to JES Stevens, head of Australian Totem Panel, April 1953, NAA: A6456, R215/016.
63 Australian Academy of Science, 'Professor John Carver (1926–2004)', transcription of interview with Bob Crompton, 1997, AAS, <www.science.org.au/learning/general-audience/history/interviews-australian-scientists/professor-john-carver-1926-2004>, accessed 9 March 2020.
64 John Carver, Letter to Ernest Titterton from Salisbury, South Australia, undated, Titterton archive, box 100, series 22, AAS.
65 Ernest Titterton, 'Preliminary Results from Group N.M. for T1', Secret memorandum to Charles Adams, AWRE, undated, Titterton archive, box 100, series 22, AAS.
66 Ernest Titterton, 'Fast Neutron Measurement at Emu on T-1 and T-2', Top-secret report, undated, Titterton archive, box 27, folder 4/89, AAS.
67 Titterton, 'Fast Neutron Measurement'.
68 Titterton, 'Fast Neutron Measurement'.
69 Titterton, 'Fast Neutron Measurement'.
70 Titterton, 'Fast Neutron Measurement'.
71 Titterton, 'Fast Neutron Measurement'.
72 Instructions to Captain Crofts for radiation survey of the north-east coast, Secret memorandum, undated and unsigned, Titterton archive, box 100, series 22, AAS.

73 Instructions to Captain Crofts.
74 JM Crofts, Reports on radiation survey on the north east coast, 31 October 1953, Titterton archive, box 100, series 22, AAS.
75 Symonds, *A History of British Atomic Tests*, p. 172.
76 William Penney, Secret letter to JES Stevens, chairman, Totem Panel, 2 November 1953, NAA: A6456, R124/018.
77 FL Westwater, commander, Royal Navy, & H Freeman, Meteorological Office, UK Air Ministry, AWRE report T12/54, Operation Totem – meteorological services, June 1954, NAA: A6455, RC246.
78 Westwater & Freeman, AWRE report.
79 'Atomic Explosion: Message from Sir Winston Churchill', Press notice no. 132/1953, 16 October 1953, NAA: A6456, R124/018.
80 Nobel Prize, 'The Nobel Prize in Literature 1953', Nobel Prize, <www.nobelprize.org/prizes/literature/1953/summary/>, accessed 27 August 2021.
81 'Other "Minor" Tests', *Sydney Morning Herald*, 28 October 1953, p. 1.
82 CA Adams, AWRE, Confidential letter to Ernest Titterton, 14 December 1953, Titterton archive, box 100, series 22, AAS.
83 Adams, Confidential letter.
84 Frederick Lindemann, Note to Churchill, 26 August 1953, Lindemann/Cherwell papers, J97, Nuffield College Library, Oxford.
85 AS Brown, Secretary, Prime Minister's Department, secret note to Prime Minister Robert Menzies, 28 September 1953, NAA: A6456, R065/003.
87 AS Brown, Secretary, Prime Minister's Department, confidential letter to JES Stevens, 29 September 1953, NAA: A6456, R065/003.
87 AS Brown, secretary, Prime Minister's Department, Secret note to Prime Minister Robert Menzies, 23 September 1953, NAA: A6456, R065/003.
88 Brown, Secret note.
89 Brown, Secret note.
90 Royal Commission, *Report*, vol. 1, p. 147.
91 Westwater & Freeman, AWRE report.
92 Royal Commission, *Report*, vol. 1.
93 Westwater & Freeman, AWRE report.
94 Winston Churchill, 'Notes on Tube Alloys, 1954', Top-secret Cabinet document, 14 December 1954, DSND 14/4, Churchill Archives Centre, University of Cambridge.
95 Heather Goodall, interview with the author, 10 June 2021.

Chapter 5 The unknowable black mist

Epigraphs: Yami Lester, Transcript of proceedings, Royal Commission into British Nuclear Tests in Australia, pp. 7080a–7540, NAA: A6448, 13; Ray Acaster, 'Sherlock Holmes and the Grim Reaper of Wallatinna: Aboriginal Oral History Solves a Scientific Mystery', *Oral History Association of Australia Journal*, no. 24, 2002, p. 22.

1 DB Waterson, 'McClelland, James Robert (1915–1999)', Biographical Dictionary of the Australian Senate: Online Edition, first published in hard copy, 2010, <biography.senate.gov.au/mcclelland-james-robert/>, accessed 24 September 2021.
2 Kingsley Palmer, 'Dealing with the Legacy of the Past: Aborigines and Atomic Testing in South Australia', *Aboriginal History*, vol. 14, no. 1/2, 1990, p. 202.

3   WB MacDougall, 'Detailed Survey of the Jangkuntjara Tribe – Their Traditional Tribal Country and Ceremonial Grounds', sent to superintendent, Woomera, 2 March 1953, Malone files.
4   MacDougall, 'Detailed Survey'.
5   James McClelland, chair, Royal Commission into British Nuclear Tests in Australia, Opening remarks before taking evidence at Marla Bore, 20 April 1985, NAA: A6448, 13, p. 7093.
6   Annette Hamilton, quoted in Royal Commission into British Nuclear Tests in Australia, *Report* (JR McClelland, president), 2 vols, AGPS, Canberra, 1985, p. 185.
7   Denis Blakeway & Sue Lloyd-Roberts, *Fields of Thunder: Testing Britain's Bomb*, George Allen & Unwin, London, 1985, p. 108.
8   Heather Goodall, 'Colonialism and Catastrophe: Contested Remembrance of Measles and Bombs in a Pitjantjatjara Community', in Paula Hamilton & Kate Darian-Smith (eds), *Memory and History in Twentieth-Century Australia*, Oxford University Press, Melbourne, 1994, p. 55.
9   Heather Goodall, 'Damage and Dispossession: Indigenous People and Nuclear Weapons on Bikini Atoll and the Pitjantjatjara Lands, 1946 to 1988', in Ann McGrath & Lynette Russell (eds), The Routledge Companion to Global Indigenous History, Routledge, London, 2021, p. 427.
10 Acaster, 'Sherlock Holmes', p. 22.
11 Acaster, 'Sherlock Holmes', p. 22.
12 Acaster, 'Sherlock Holmes', p. 22.
13 JA Cooke, on behalf of acting regional manager, Department of Aboriginal Affairs South Australian Region, 'Aboriginals and Atomic Tests', Letter to secretary, Department of Aboriginal Affairs, Canberra, 4 September 1980, Malone files.
14 Cooke, 'Aboriginals and Atomic Tests'.
15 Heather Goodall, interview with the author, 10 June 2021.
16 Aboriginal Health Unit, South Australian Health Commission, *A Survey of Diseases That May Be Related to Radiation among Pitjantjatjara on Remote Reserves*, Adelaide, 18 February 1981, p. 2; Robert Ball, David English & Peter de Ionno, 'A "Black Mist" that Brought Death', The Advertiser (Adelaide), 3 May 1980, p. 1.
17 Aboriginal Health Unit, *A Survey of Diseases*, p. 3.
18 Aboriginal Health Unit, *A Survey of Diseases*, p. 3.
19 Aboriginal Health Unit, *A Survey of Diseases*, p. 15.
20 Aboriginal Health Unit, *A Survey of Diseases*, p. 15.
21 Aboriginal Health Unit, *A Survey of Diseases*, p. 15.
22 Expert Committee on the Review of Data on Atmospheric Fallout Arising from British Nuclear Tests in Australia, *Report of the Expert Committee on the Review of Data on Atmospheric Fallout Arising from British Nuclear Tests in Australia* (CB Kerr, chairman; Kerr report), AGPS, Canberra, 31 May 1984.
23 Geoffrey Eames, counsel for Aboriginal people, Opening remarks at Marla Bore, 20 April 1985, Transcript of proceedings, Royal Commission into British Nuclear Tests in Australia, NAA: A6448, 13, p. 7088.
24 Andrew Collett, correspondence with the author, 6 December 2021.
25 William Penney, Transcript of proceedings, Royal Commission into British Nuclear Tests in Australia, NAA: A6448, 9, p. 4348.
26 Penney, Transcript of proceedings, p. 4348.

27   Lorna Arnold & Mark Smith, *Britain, Australia and the Bomb: The Nuclear Tests and Their Aftermath*, 2nd edn, Palgrave Macmillan, New York, 2006, p. 71.
28   Arnold & Smith, *Britain, Australia and the Bomb*, p. 71.
29   Royal Commission, *Report*, vol. 2, p. 599.
30   GA Williams, RS O'Brien, M Grzechnik & KN Wise, 'Estimates of Radiation Doses to the Skin for People Camped at Wallatinna during the UK Totem 1 Atomic Weapons Test', *Radiation Protection Dosimetry*, vol. 174, no. 3, 2017, p. 323.
31   Williams, O'Brien, Grzechnik & Wise, 'Estimates of Radiation Doses', p. 323.
32   Williams, O'Brien, Grzechnik & Wise, 'Estimates of Radiation Doses', p. 322.
33   Williams, O'Brien, Grzechnik & Wise, 'Estimates of Radiation Doses', p. 323.
34   Williams, O'Brien, Grzechnik & Wise, 'Estimates of Radiation Doses', pp. 324–25.
35   Williams, O'Brien, Grzechnik & Wise, 'Estimates of Radiation Doses', p. 325.
36   Williams, O'Brien, Grzechnik & Wise, 'Estimates of Radiation Doses', p. 326.
37   Williams, O'Brien, Grzechnik & Wise, 'Estimates of Radiation Doses', p. 331.
38   Environmental Protection Agency, 'Radionuclide Basics: Cobalt-60', EPA, <www.epa.gov/radiation/radionuclide-basics-cobalt-60>, accessed 4 July 2021.
39   Williams, O'Brien, Grzechnik & Wise, 'Estimates of Radiation Doses', p. 329.
40   Williams, O'Brien, Grzechnik & Wise, 'Estimates of Radiation Doses', p. 330.
41   Royal Commission, *Report*, vol. 1, p. 185.
42   Royal Commission, *Report*, vol. 1, p. 185.
43   Jun Takada, Masaharu Hoshi, Shozo Sawada & Masanobu Sakanoue, 'Uranium Isotopes in Hiroshima "Black Rain" Soil', *Journal of Radiation Research*, vol. 24, 1983, p. 229.
44   Yami Lester, *Autobiography of Yami Lester*, 2nd edn, Jukurrpa Books, Alice Springs, 2000, p. 31.
45   Lester, *Autobiography*, p. 32.
46   Lester, *Autobiography*, p. 32.
47   David Tonkin, Transcript of proceedings, Royal Commission into British Nuclear Tests in Australia, NAA: A6448, 15, p. 8632a.
48   Lester, *Autobiography*, p. 187.
49   Lester, *Autobiography*, p. 187.
50   Lester, *Autobiography*, p. 187.
51   Lester, *Autobiography*, p. 188.
52   Lester, *Autobiography*, p. 189.
53   Lester, *Autobiography*, p. 189.
54   David Leigh & Paul Lashmar, 'Forgotten Victims of the "Rolling Black Mist"', originally published in *Observer*, edited and republished in *Bulletin*, vol. 103, no. 5360, 12 April 1984, pp. 97–100.
55   Lester, *Autobiography*, p. 196.
56   Lester, *Autobiography*, p. 195.
57   Kanytji & Pingkayi, Transcript of proceedings, Royal Commission into British Nuclear Tests in Australia, NAA: A6448, 13, pp. 7189–90.
58   Lallie Lennon, Transcript of proceedings, Royal Commission into British Nuclear Tests in Australia, NAA: A6448, 13, p. 7147.
59   Lennon, Transcript of proceedings, p. 7154.
60   Lennon, Transcript of proceedings, p. 7154.
61   Ellen Giles, Statements from Australian witnesses: E–J, Royal Commission into British Nuclear Tests in Australia, NAA: A6450, 2.

62 Lester, *Autobiography*, p. 194.
63 Almerta Lander, Transcript of proceedings, Royal Commission into British Nuclear Tests in Australia, NAA: A6448, 13, p. 7098.
64 Lander, Transcript of proceedings, p. 7102.
65 Lander, Transcript of proceedings, p. 7106.
66 Richard Bradshaw, former Pitjantjatjara Council lawyer, interview with the author, 25 May 2021.
67 Bradshaw, interview.
68 Heather Goodall, '"The Whole Truth and Nothing But ..."': Some Intersections of Western Law, Aboriginal History and Community Memory', *Journal of Australian Studies*, no. 32, March 1992, p. 104.
69 Goodall, 'The Whole Truth and Nothing But ...', p. 115.
70 Goodall, 'The Whole Truth and Nothing But ...', p. 116.
71 Goodall, interview.
72 Goodall, interview.
73 Goodall, 'The Whole Truth and Nothing But ...', p. 115.
74 MacDougall, 'Detailed Survey'.
75 Goodall, 'The Whole Truth and Nothing But ...', p. 117.
76 Goodall, interview.
77 Goodall, interview.
78 Geoffrey Eames, counsel for Aboriginal people, Opening remarks on the first day of hearings in London, Transcript of proceedings, Royal Commission into British Nuclear Tests in Australia, NAA: A6448, 9, p. 12.

Chapter 6 Secrets and safety lies
Epigraphs: DE Barnes, GC Dale, RA Siddons & ER Woodcock, High Explosives Research report no. A32 – Materials and Physical Research Division (HER) – Airborne Contamination at the Totem Trial, UK Ministry of Supply, May 1953, NAA: A6454, ZB3, p. 3; Official guide for visitors to X200, 1953, NAA: A6456, R233/001; Ronald Monson, 'Security Was Tight on Project X200', *Daily Telegraph (Sydney)*, 24 October 1953, p. 21.
1 Ben Cockram, Top-secret letter to Robert Menzies, 3 March 1953, NAA: A6456, R124/018.
2 DE Barnes, GC Dale, RA Siddons & ER Woodcock, High Explosives Research report.
3 DE Barnes, GC Dale, RA Siddons & ER Woodcock, High Explosives Research report, p. 19.
4 DE Barnes, GC Dale, RA Siddons & ER Woodcock, High Explosives Research report, p. 19.
5 DE Barnes, GC Dale, RA Siddons & ER Woodcock, High Explosives Research report, p. 21.
6 Geoffrey Eames & Andrew Collett, Final submission on behalf of Aboriginal groups and individuals, Royal Commission into British Nuclear Tests in Australia, 16 September 1985, NAA: A6455, RC862, p. 119.
7 Eames & Collett, Final submission, p. 121.
8 William Penney, Transcript of proceedings, Royal Commission into British Nuclear Tests in Australia, NAA: A6448,12, p. 7052.
9 William Penney, Transcript of proceedings, Royal Commission into British Nuclear Tests in Australia, NAA: A6448, 9, pp. 4339–4340.

10 'Atomic Tests "Threatened No-One"', *Canberra Times*, 15 May 1985, p. 3.
11 Cover letter addressed to minister for science, 12 March 1980, AIRAC report on the Emu atomic test sites, Department of Environment, Housing and Community Development, Atomic Energy Commission, Report on residual radioactivity at Emu Field, NAA: A6456, R057/003.
12 Lorna Arnold & Mark Smith, *Britain, Australia and the Bomb: The Nuclear Tests and Their Aftermath*, 2nd edn, Palgrave Macmillan, New York, 2006, p. 65.
13 Ernest Titterton, 'The Role of "Observers" and the AWTSC at British Weapons Tests, in Australia', 19 May 1985, Titterton archive, box 99, series 22, AAS.
14 Titterton, 'The Role of "Observers"'.
15 Minutes of the seventh meeting of interdepartmental [Totem] panel, Top-secret, Melbourne, 24 June 1953, NAA: A6456, R095/012.
16 Leslie Martin, University of Melbourne, Top-secret letter to Prime Minister Robert Menzies, 17 June 1953, NAA: A6456, R095/012.
17 Martin, Top-secret letter.
18 Royal Commission into British Nuclear Tests in Australia, *Report* (JR McClelland, president), 2 vols, AGPS, Canberra, 1985, p. 144.
19 Royal Commission, *Report*, p. 144.
20 JES Stevens, Top-secret teleprinter message to Elmhirst, 15 May 1953, DEFE 16/523, NAUK.
21 Martin, Top-secret letter.
22 Royal Commission into British Nuclear Tests in Australia, *Report: Conclusions and Recommendations* (JR McClelland, J Fitch, WJA Jones), AGPS, Canberra, 1985, p. 8.
23 Michael S Goodman, *Spying on the Nuclear Bear: Anglo-American Intelligence and the Soviet Bomb*, Stanford University Press, Stanford, 2007, p. 97.
24 Goodman, *Spying on the Nuclear Bear*, p. 97.
25 Arnold & Smith, *Britain, Australia and the Bomb*, p. 55.
26 Arnold & Smith, *Britain, Australia and the Bomb*, p. 55.
27 Royal Commission, *Report*, pp. 142, 151.
28 Arnold & Smith, *Britain, Australia and the Bomb*, p. 66.
29 FC How, UK Ministry of Supply, Official Committee on Atomic Energy, Top-secret briefing document on Operation Totem, 2 February 1953, AB 16/1743, NAUK.
30 'Welcome to the Claypan', An introduction to living and working at the Emu claypan campsite, 1953, NAA: D2861, WELCOME TO THE CLAYPAN.
31 'Welcome to the Claypan', p. 6.
32 J McEnhill, UK Ministry of Supply, Operation Totem summary plan, part 1, NAA: A6454, ZB1.
33 McEnhill, Operation Totem summary plan.
34 How, Top-secret briefing document.
35 How, Top-secret briefing document.
36 How, Top-secret briefing document.
37 How, Top-secret briefing document.
38 How, Top-secret briefing document.
39 Royal Commission, *Report*, vol. 1, p. 117.
40 Commonwealth Relations Office, Telegram to UK high commissioner in New Zealand, [1953], AB 16/1743, NAUK.

41 Erron Toon, national secretary, Maralinga and Monte Bello Atomic Ex-Servicemen's Association, Letter to Bill Hayden, member for Oxley, 18 May 1984, Malone files.
42 Erron Toon, Transcript of proceedings, Royal Commission into British Nuclear Tests in Australia, NAA: A6448, 2, p. 377.
43 RAAF, Service conduct sheet, David Richard Bailey, 16 September 1956, NAA: A12372, A21830, BAILEY, David Richard, Service record.
44 RAAF, Assessment of conduct and trade proficiency for David Bailey, 9 September 1956, NAA: A12372, A21830, BAILEY, David Richard, Service record.
45 RAAF, Assessment.
46 Stephen Hogeveen, interview with the author, 15 June 2021.
47 David Bailey, Handwritten note to officer in charge, Defence Records, 18 April 1997, NAA: A12372, A21830, BAILEY, David Richard, Service record.
48 Hogeveen, interview.
49 Hogeveen, interview.
50 Hogeveen, interview.
51 Hogeveen, interview.
52 Goodall, interview with the author, 10 June 2021.
53 Goodall, interview.
54 Peter McClellan, 'Who is telling the truth? Psychology, common sense and the law', *Australian Law Journal*, 2006, vol. 80 (10), p.655–666, p. 656.
55 Mark Earner, Transcript of proceedings, Royal Commission into British Nuclear Tests in Australia, October 1984, NAA: A6448, 2, p. 508.
56 Earner, Transcript of proceedings, p. 513.
57 Earner, Transcript of proceedings, p. 514.
58 Royal Commission, *Report*, vol. 2, p. 434.
59 McClellan, 'Who Is Telling the Truth?', p. 2.
60 McClellan, 'Who Is Telling the Truth?', p. 1.

Chapter 7 Flying through the clouds
Epigraphs: Denis Wilson & GH Dhenin, 'Some Aspects of Aviation Medicine in Regard to Radiological Hazards', *Proceedings of the Royal Society of Medicine*, vol. 48, no. 1, 7 October 1954, p. 4; Roy Cosgrove, Statements from Australian witnesses: A–D, Royal Commission into British Nuclear Tests in Australia, NAA: A6450, 1.
1 Operation Totem – Emu Field 1953, Details of RAAF participation in British atomic tests held in Australia 1952–1957, NAA: A6456, R021/001.
2 Royal Commission into British Nuclear Tests in Australia, *Report* (JR McClelland, president), 2 vols, AGPS, Canberra, 1985, p. 215.
3 Minutes of the seventh meeting of interdepartmental [Totem] panel, Top-secret, Melbourne, 24 June 1953, NAA: A6456, R095/012.
4 Minutes of the seventh meeting.
5 William Penney, Transcript of proceedings, Royal Commission into British Nuclear Tests in Australia, NAA: A6448, 9, p. 4305.
6 Penney, Transcript of proceedings, p. 4305.
7 Penney, Transcript of proceedings, p. 4305.

8   Expert Committee on the Review of Data on Atmospheric Fallout Arising from British Nuclear Tests in Australia, *Report of the Expert Committee on the Review of Data on Atmospheric Fallout Arising from British Nuclear Tests in Australia* (CB Kerr, chairman; Kerr report), AGPS, Canberra, 31 May 1984, p. 17.
9   Expert Committee, *Report*, p. 17.
10 JL Symonds, *A History of British Atomic Tests in Australia*, AGPS, Canberra, 1985, pp. 210–11.
11 Expert Committee, *Report*, p. 17.
12 Rex Naggs, Transcript of proceedings, Royal Commission into British Nuclear Tests in Australia, NAA: A6448, 4, p. 1265.
13 Naggs, Transcript of proceedings, p. 1267.
14 Royal Commission, *Report*, p. 140.
15 Royal Commission, *Report*, p. 140.
16 RAAF operational information, Operation Totem, No. 6 (B) Squadron, NAA: A11288, 28/7/AIR PART 1.
17 Symonds, *A History of British Atomic Tests*, p. 166.
18 Symonds, *A History of British Atomic Tests*, p. 167.
19 Symonds, *A History of British Atomic Tests*, p. 167.
20 Cosgrove, Statements from Australian witnesses.
21 RAAF, No. 82 (B) Wing, Operation order no. 3/53, Operation 'Totem', NAA: A11288, 28/7/AIR PART 1.
22 Royal Commission, *Report*, vol. 1, p. 228.
23 Group Captain DW Colquhoun, officer commanding, Secret letter to director, Air Force Intelligence, Melbourne, 19 November 1953, NAA: A13064, DPIE 92/001390.
24 Colquhoun, Secret letter.
25 Royal Commission, *Report*, p. 418.
26 David Colquhoun, Transcript of proceedings, Royal Commission into British Nuclear Tests in Australia, NAA: A6448, 15.
27 Royal Commission, *Report*, p. 140.
28 Symonds, *A History of British Atomic Tests*, p. 113.
29 Colquhoun, quoted in Symonds, *A History of British Atomic Tests*, p. 170.
30 Colquhoun, quoted in Symonds, *A History of British Atomic Tests*, p. 170.
31 Colquhoun, quoted in Symonds, *A History of British Atomic Tests*, p. 170.
32 'Crews "Over-Exposed" and "Under-Protected" in A-Test', *Canberra Times*, 10 October 1984, p. 12.
33 Lancelot Edwards, Statements from Australian witnesses: E–J, Royal Commission into British Nuclear Tests in Australia, Submission to the Royal Commission, NAA: A6450, 2, p. 2.
34 Edwards, Statements from Australian witnesses, p. 3.
35 J Austin, AWRE, Operation Totem team report, Team RH5 – decontamination, The prevention and removal of radio-active contamination, part VI, Decontamination of aircraft and health control at Woomera and Amberley, 6 September 1954, NAA: A6454, Z100.
36 Austin, Operation Totem team report.
37 Austin, Operation Totem team report.
38 Austin, Operation Totem team report.

39 Group Captain DA Wilson, Report no. T3/54, Canberra flight report, Operation Hot Box, ES 5/5, NAUK.
40 Wilson, Report.
41 Atomic Weapons Research Establishment, *Operation Totem*, film, Aldermaston, 1953, available at International War Museums, <www.iwm.org.uk/collections/item/object/1060036296>, accessed 2 February 2021.
42 Lorna Arnold & Mark Smith, *Britain, Australia and the Bomb: The Nuclear Tests and Their Aftermath*, 2nd edn, Palgrave Macmillan, New York, 2006, p. 61.
43 RAAF Eastern Area Detachment Woomera, Operation order no. 1/53, RAAF operational information.
44 Operation 'Totem' consolidated report, RAAF, Headquarters Home Command, Penrith, 5 March 1954, NAA: A6455, RC36, p. 13.
45 Wilson & Dhenin, 'Some Aspects of Aviation Medicine', p. 3.
46 Wilson, Report.
47 GA Williams, radiation scientist, correspondence with the author, 30 July 2021.
48 Wilson & Dhenin, 'Some Aspects of Aviation Medicine', p. 3.
49 RA Siddons, AWRE report no. E1/54, The rise of the cloud produced in an atomic explosion, Aldermaston, October 1954, ES 3/6, NAUK, p. 15.
50 Wilson, Report.
51 Wilson, Report.
52 Wilson, Report.
53 Wilson, Report.
54 Symonds, *A History of British Atomic Tests*, p. 166.
55 Wilson, Report.
56 Wilson, Report.
57 Wilson & Dhenin, 'Some Aspects of Aviation Medicine', p. 2.
58 Wilson & Dhenin, 'Some Aspects of Aviation Medicine', p. 2.
59 Wilson & Dhenin, 'Some Aspects of Aviation Medicine', p. 3.
60 Wilson & Dhenin, 'Some Aspects of Aviation Medicine', p. 3.
61 Wilson & Dhenin, 'Some Aspects of Aviation Medicine', p. 3.
62 Morris Newman, Statements from Australian witnesses: K–O, Royal Commission into British Nuclear Tests in Australia, NAA: A6450, 3.
63 Newman, Statements from Australian Witnesses.
64 Morris Newman, Death certificate, Registration number 2000/60765, Queensland Government, 2000, provided by Michele Newman.
65 Commonwealth Department of Veterans' Affairs, Letter to Morris Newman, 13 June 1996, provided by Michele Newman.
66 Michele Newman, correspondence.
67 Charles Geschke, Statements from Australian witnesses: E–J, Royal Commission into British Nuclear Tests in Australia, NAA: A6450, 2.
68 Eastern Area Detachment Woomera, Restricted teletype message to 82 Wing, 28 October 1953, NAA: A13064, DPIE 92/001390.
69 EA Daley, air vice marshal, director general, Medical Services, RAAF, Secret minute paper, 3 December 1953, Department of Air, Radiological health during Operation 'Hurricane' Monte Bello Islands, October/November 1952, and Operation 'Totem', Emu claypan, October/November 1953, NAA: A6456, R021/001 PART 79.
70 Daley, Secret minute paper.
71 Daley, Secret minute paper.

72 Richard Nettley, retired RAF, Statements from UK witnesses: M–Z, Royal Commission into British Nuclear Tests in Australia, NAA: A6449, 2.
73 EA Daley, air vice marshal, director general, Medical Services, RAAF, Secret minute, 25 May 1954, Department of Air, Medical implications in the RAAF, NAA: A6456, R024/001.
74 FL Westwater, commander, Royal Navy, & H Freeman, Meteorological Office, UK Air Ministry, AWRE report T12/54, Operation Totem – meteorological services, June 1954, NAA: A6455, RC246.
75 Royal Commission, *Report*, p. 204.
76 Symonds, *A History of British Atomic Tests*, p. 171.
77 Symonds, *A History of British Atomic Tests*, p. 171.
78 Symonds, *A History of British Atomic Tests*, p. 171.
79 Symonds, *A History of British Atomic Tests*, p. 173.
80 Allen Clark, Statements from Australian witnesses: A–D, Royal Commission into British Nuclear Tests in Australia, NAA: A6450, 1.
81 Cosgrove, Statements from Australian witnesses.
82 Cosgrove, Statements from Australian witnesses.
83 Royal Commission, *Report*, vol. 1, p. 228.
84 John Clarke QC, Air Marshal Doug Riding & Dr David Rosalky 'British Atomic Tests', in Department of Veterans' Affairs, *Report of the Review of Veterans' Entitlements*, vol. 2, Commonwealth of Australia, Canberra, 6 January 2003, p. 391.
85 Clarke, 'British Atomic Tests', p. 394.
86 Department of Veterans' Affairs, 'Support for Participants in British Nuclear Testing: Compensation and Health Care Treatment under the Veterans' Entitlements Act 1986', DVA, last updated 19 November 2020, <www.dva.gov.au/financial-support/compensation-claims/claims-nuclear-test-or-bomb-site-exposure/support-participants>, accessed 8 August 2021.

Chapter 8 The people's witness
Epigraph: James Cameron, 'When the Desert Skies Caught Fire', *Age*, 16 October 1953, p. 2.
1 Frederick Lindemann, Note to Churchill, 26 August 1953, Lindemann/Cherwell papers, J97, Nuffield College Library, Oxford.
2 Lindemann, Note.
3 Lindemann, Note.
4 Royal Commission into British Nuclear Tests in Australia, *Report: Conclusions and Recommendations* (JR McClelland, J Fitch, WJA Jones), AGPS, Canberra, 1985, p. 8.
5 Ben Cockram, Top-secret letter to Robert Menzies, 3 March 1953, NAA: A6456, R124/018.
6 AD McKnight, Prime Minister's Department, Top-secret personal letter to Ben Cockram, UK High Commission to Australia, 23 January 1953, NAA: A6456, R124/018.
7 McKnight, Top-secret personal letter.
8 McKnight, Top-secret personal letter.
9 Totem Panel, Minutes, 24 July 1953, NAA: A6456, R095/012.
10 Totem Panel, Minutes.
11 Minutes of the seventh meeting of interdepartmental [Totem] panel, Top-secret, Melbourne, 24 June 1953, NAA: A6456, R095/012.

12   Harold Dvoretsky, 'Choice of a Weapon: Woomera Test Experts in Big Decision', Sydney *Sunday Telegraph*, 16 August 1953, Atomic test at Woomera, NAA: A5954, 2167/1.
13   Dvoretsky, 'Choice of a Weapon'.
14   Dvoretsky, 'Choice of a Weapon'.
15   'Woomera's Big Role', *Herald* (Melbourne), 3 August 1953, Atomic test at Woomera.
16   AD McKnight, Prime Minister's Department, Letter to George Davey, office of the High Commissioner for the United Kingdom, Canberra, 30 September 1952, NAA: A6456, R096/006.
17   McKnight, Letter.
18   Frederick Shedden, secretary, Department of Defence, Top-secret letter to AD McKnight, Prime Minister's Department, 15 September 1952, NAA: A6456, R096/006.
19   Tim Sherratt, 'A Political Inconvenience: Australian Scientists at the British Atomic Weapons Tests, 1952–53', *Historical Records of Australian Science*, vol. 6, no. 2, December 1985, p. 143.
20   McKnight, Letter.
21   'America Can Watch This – "If…": Secret Atom Missiles Fly Here for October Test', *Argus*, 3 August 1953, Atomic test at Woomera.
22   'America Can Watch This'.
23   'Woomera Atom Test in October: Sir William Penney Coming Again', *Age*, 3 August 1953, Atomic test at Woomera.
24   Trevor Smith, 'A Test Here in October: U.K. Minister for Woomera', *Herald* (Melbourne), 25 July 1953, Atomic test at Woomera.
25   Smith, 'A Test Here in October'.
26   Herald Special Service, 'A-Bombs Flown Here', *Herald* (Melbourne), 22 July 1953, Atomic test at Woomera.
27   'New A-Bomb for Woomera', Sydney *Sunday Telegraph*, 21 July 1953, Atomic test at Woomera.
28   An account of the *Daily Mirror*'s populism, and its fierce rivalry with the *Sun*, may be found in Sandra Hall, *Tabloid Man: The Life and Times of Ezra Norton*, HarperCollins, Sydney, 2008. See in particular p. 1.
29   'His Responsibility!', *Daily Mirror* (Sydney), 2 October 1953, p. 2.
30   Staff correspondent, 'Public Safety and the Woomera Tests', *Sydney Morning Herald*, 9 October 1953, p. 2.
31   Staff correspondent, 'Public Safety and the Woomera Tests'.
32   Staff correspondent, 'Public Safety and the Woomera Tests'.
33   Staff correspondent, 'Long-Range Risks in Atomic Tests?', *Sydney Morning Herald*, 10 October 1953, p. 2.
34   Robert Menzies, 'Australia Today', draft, *Man to Man*, radio program, broadcast no. 11, 21 October 1953, quoted in Erin Ison, 'Bombs, "Reds under the Bed" and the Media: The Menzies Government's Manipulation of Public Opinion, 1949–1957', *Humanity* (University of Newcastle), 2008.
35   'Unofficial Background Scientific Information for Press Observers', undated, Titterton archive, box 100, series 22, AAS.
36   'Unofficial Background Scientific Information'.
37   JL Symonds, *A History of British Atomic Tests in Australia*, AGPS, Canberra, 1985, p. 161.

38 Ronald Monson, 'Security Was Tight on Project X200', *Daily Telegraph (Sydney)*, 24 October 1953, p. 21.
39 Monson, 'Security Was Tight'.
40 A 'Herald' special reporter who watched the atomic explosion on the Woomera range from 15 miles away, 'Atom Explosion Success: Dawn Blast at Woomera', *Sydney Morning Herald*, 16 October 1953, p. 1.
41 A 'Herald' special reporter, 'Atom Explosion Success', p. 1.
42 Special correspondent in London, 'How an Atomic Bomb Explodes', *Sydney Morning Herald*, 16 October 1953, p. 2.
43 'Months of Hard, Lonely Work Paved Way for Atomic Explosion', *Sydney Morning Herald*, 16 October 1953, p. 4.
44 'Months of Hard, Lonely Work'.
45 'Months of Hard, Lonely Work'.
46 'Second A-Blast Successful: Trial Series Ends', *Sydney Morning Herald*, 28 October 1953, p. 1.
47 'Other "Minor" Tests', *Sydney Morning Herald*, 28 October 1953, p. 1.
48 Cameron, 'When the Desert Skies Caught Fire', p. 2.
49 Cameron, 'When the Desert Skies Caught Fire'.
50 Cameron, 'When the Desert Skies Caught Fire'.
51 Cameron, 'When the Desert Skies Caught Fire'.
52 AE Buchanan, '"D" Notice No. 10', Defence, Press and Broadcasting Committee, NAA: A1209, 1957/5486.
53 Arthur Fadden, Letter to press chiefs, UK Atomic Test Woomera – D-notice to the press, quoted in Symonds, *A History of British Atomic Tests*, p. 151.
54 Ben Cockram, Top-secret letter to Neil Pritchard, 19 January 1953, AB 16/1743, NAUK.
55 'No Danger from Bomb Test: P.M. Assures of Safety', *Age*, 9 September 1953, Atomic test at Woomera.
56 'No Danger from Bomb Test'.
57 'US Reps at Atom Tests', *Sunday Sun & Guardian* (Sydney), 6 September 1953, Atomic test at Woomera.
58 'US Reps at Atom Tests'.
59 JES Stevens, Top-secret letter to Allan McKnight, 7 May 1953, NAA: 6456, R215/016.
60 Stevens, Top-secret letter.
61 AS Brown, secretary, Prime Minister's Department, Secret note to Prime Minister Robert Menzies, 28 September 1953, NAA: A6456, R065/003.
62 William Penney, Restricted de-briefing notice for UK staff, undated, Titterton archive, box 100, series 22, AAS.
63 Penney, Restricted de-briefing notice.
64 'Atomic Bombs, Uranium and Australia's Future', *Sydney Morning Herald*, 8 October 1953, p. 2.
65 'Atomic Bombs, Uranium and Australia's Future'.
66 'Atomic Bombs, Uranium and Australia's Future'.
67 Monson, 'Security Was Tight'.
68 'Atom Bombs in Our Arid Lands', *Sunday Herald* (Sydney), 4 October 1953, p. 2.

Chapter 9 Australia's atomic bargaining chip

Epigraph: Howard Beale, minister for supply, Top-secret letter to Percy Spender, Australian ambassador to Washington, 22 October 1954, NAA: A6456, R065/011.

1. JES Stevens, chairman, Australian Atomic Energy Commission, Secret report prepared for the minister for external affairs, Notes on proposal for an international atomic energy pool as it affects Australia, 21 April 1954, NAA: A6456, R065/011.
2. Alice Cawte, *Atomic Australia*, New South Wales University Press, Sydney, 1992, p. 4.
3. Darren Holden, 'Mark Oliphant and the Invisible College of the Peaceful Atom', PhD thesis, University of Notre Dame Australia, 2019, p. 119.
4. Holden, 'Mark Oliphant', p. 100.
5. Stevens, Secret report.
6. Cawte, *Atomic Australia*, pp. 4–5.
7. Cabinet Committee on Uranium, Programme for atomic development, Agendum no. UC/8, May 1953, NAA: A4940, C1730 PART 1.
8. Cawte, *Atomic Australia*, p. 70.
9. Secret departmental briefing note, Notes furnished to the Australian ambassador in Washington on the bilateral agreement between Australia and the U.S.A., [probably 1954], NAA: A6456, R065/011; LJ Butler, 'The Central African Federation and Britain's Post-War Nuclear Programme: Reconsidering the Connections', *Journal of Imperial and Commonwealth History*, vol. 36, no. 3, September 2008, p. 517.
10. JL Symonds, *A History of British Atomic Tests in Australia*, AGPS, Canberra, 1985, p. 4.
11. Butler, 'The Central African Federation', p. 517.
12. James Chadwick, Top-secret letter to Lord Portal, 'Private and Personal', 8 April 1946. Lord Portal Archive Atomic Energy folder B2, UK Military Atomic Energy Chronology 1945–1969 (including V bomber force). Portal Collection, Christ Church Library, Oxford.
13. Margaret Gowing with Lorna Arnold, *Independence and Deterrence: Britain and Atomic Energy, 1945–1952*, UK Atomic Energy Authority, St Martin's Press, New York, 1974, p. 349.
14. UK House of Commons, Debates, 9 November 1953, vol. 520, col. 588 (Duncan Sandys).
15. Defence Committee, Agendum no. 3/1950, Secret notes, Atomic warfare resources, NAA: A5954, 2330/2.
16. Defence Committee, Agendum.
17. Cawte, *Atomic Australia*, p. 66.
18. Frederick Lindemann, Note to Churchill, 29 December 1952, Lindemann/Cherwell papers, J122, Nuffield College Library, Oxford.
19. Butler, 'The Central African Federation', p. 511.
20. Lindemann, Note, 29 December 1952.
21. Lindemann, Note, 29 December 1952.
22. Lindemann, Note, 29 December 1952.
23. Lindemann, Note, 29 December 1952.
24. Lindemann, Note, 29 December 1952.
25. Frederick Lindemann, Note to Churchill, 4 May 1953, Lindemann/Cherwell papers, J138, Nuffield College Library, Oxford.

26 Lindemann, Note, 4 May 1953.
27 Butler, 'The Central African Federation', p. 516.
28 'Discussions with Mr. Gordon Dean and Mr. Lewis Strauss, the Prime Minister and Mr. Dean', Secret departmental briefing note, undated, NAA: A6456, R065/011.
29 Lorna Arnold & Mark Smith, *Britain, Australia and the Bomb: The Nuclear Tests and Their Aftermath*, 2nd edn, Palgrave Macmillan, New York, 2006, p. 23.
30 Howard Beale, minister for supply, Top-secret letter to Robert Menzies, prime minister, 28 May 1953, NAA: A4940, C1730 PART 1.
31 Beale, Top-secret letter.
32 Beale, Top-secret letter.
33 Graham Farmelo, *Churchill's Bomb: A Hidden History of Science, War and Politics*, Faber & Faber, London, 2013, p. 318.
34 Lewis Betts, *Duncan Sandys and British Nuclear Policy-Making*, Palgrave Macmillan, London, 2016, p. 24.
35 'Talks on Atomic Weapons: Ministers Meet in Canberra', *Daily Telegraph (Sydney)*, 31 August 1953, Atomic test at Woomera, NAA: A5954, 2167/1.
36 Duncan Sandys, Letter to William Penney via Ministry of Supply, 12 September 1953, NAA: A1533, 1954/299.
37 'US Reps at Atom Tests', *Sunday Sun & Guardian* (Sydney), 6 September 1953, Atomic test at Woomera.
38 'Woomera Trials in October', *Herald* (Melbourne), 4 September 1953, Atomic test at Woomera.
39 Frederick Lindemann, Note to Churchill, 4 June 1953, Lindemann/Cherwell papers, J97, Nuffield College Library, Oxford.
40 Cawte, *Atomic Australia*, p. 57.
41 Frederick Lindemann, Note to Churchill, 23 June 1953, Lindemann/Cherwell papers, J97, Nuffield College Library, Oxford.
42 Martin Theaker, 'Being Nuclear on a Budget: Churchill, Britain and "Atoms for Peace" 1953–1955', *Diplomacy & Statecraft*, vol. 27, no. 4, 2016, p. 647.
43 Cawte, *Atomic Australia*, p. 58.
44 Symonds, *A History of British Atomic Tests*, p. 136.
45 Philip Cunliffe-Lister, secretary of state for Commonwealth relations, Private note to Robert Menzies, 1 July 1953, NAA: A4949, C2047.
46 Cunliffe-Lister, Private note.
47 Cawte, *Atomic Australia*, p. 58.
48 D Gilfedder, Australian Atomic Energy Commission, Exchange of information – Australia and the United Kingdom, 31 December 1954, Briefing notes for the prime minister's visit to UK and USA, 1955, NAA: A6456, R065/011.
49 'Our Future Lies in This Asset', editorial, *Argus*, 8 October 1953, p. 2.
50 'The Price of Uranium', plain English editorial, *Bulletin*, vol. 74, no. 3843, 7 October 1953. p. 6.
51 Cabinet Committee on Uranium, Top-secret minutes, Attached draft notes in response to meeting with Lord Cherwell, 6 October 1953, NAA: A4949, C2047.
52 AS Brown, secretary, Prime Minister's Department, Top-secret note to Robert Menzies, 6 October 1953, NAA: A4949, C2047.
53 *Daily Mirror (Sydney)*, quoted in Cawte, *Atomic Australia*, 1992, p. 59.
54 UK Cabinet Office, Top-secret message to Lord Cherwell's office, 13 October 1953, NAUK, EG 1/35.

55 UK Cabinet Office, Top-secret message.
56 Robert Menzies, Personal letter to Winston Churchill, 9 November 1953, EG 1/35, NAUK.
57 Menzies, Personal letter.
58 Menzies, Personal letter.
59 Menzies, Personal letter.
60 Robert Menzies, Top-secret personal letter to marquis of Salisbury, 17 March 1954, NAA: A6706, 50.
61 AS Brown, secretary, Prime Minister's Department, Top-secret note to Robert Menzies, 9 March 1954, NAA: A6706, 50.
62 D Gilfedder, Australian Atomic Energy Commission, Uranium production in Australia, 31 December 1954, Briefing notes for the Prime Minister's visit.
63 Gilfedder.
64 Neil Pritchard, Top-secret letter to Ben Cockram, 25 November 1953, FCO 1/6, NAUK.
65 Pritchard, Top-secret letter.
66 Wayne Reynolds, 'Rethinking the Joint Project: Australia's Bid for Nuclear Weapons, 1945–1960', *Historical Journal*, vol. 41, no. 3, 1998, p. 863.
67 Pritchard, Top-secret letter.
68 Jim Walsh, 'Surprise Down Under: The Secret History of Australia's Nuclear Ambitions', *Nonproliferation Review*, Autumn 1997, p. 4.
69 Reynolds, 'Rethinking the Joint Project', p. 863.
70 Reynolds, 'Rethinking the Joint Project', p. 863.
71 Frederick Lindemann, Draft reply to question from Edelman, 10 November 1953, Lindemann/Cherwell papers, J97, Nuffield College Library, Oxford.
72 Lindemann, Draft reply.
73 Menzies, Personal letter.
74 Menzies, Personal letter.
75 United States Offices of the Joint Committee on Atomic Energy, Media release, 20 December 1954, NAA: A5462, 138/2/1.
76 United States Offices, Media release.
77 John Krige, 'The Peaceful Atom as Political Weapon: Euratom and American Foreign Policy in the Late 1950s', *Historical Studies in the Natural Sciences*, vol. 38, no. 1, Winter 2008, p. 21.
78 Krige, 'The Peaceful Atom', p. 43.
79 Dwight D Eisenhower, 'Address before the General Assembly of the United Nations on Peaceful Uses of Atomic Energy, New York City', 8 December 1953, transcription at American Presidency Project, <www.presidency.ucsb.edu/documents/address-before-the-general-assembly-the-united-nations-peaceful-uses-atomic-energy-new>, accessed 28 August 2021.
80 Matthew Fuhrmann, 'Spreading Temptation: Proliferation and Peaceful Nuclear Cooperation Agreements', *International Security*, vol. 34, no. 1, Summer 2009, p. 49.
81 Cawte, *Atomic Australia*, p. 61.
82 Robin Gerster, 'Down the Yellowcake Road: The Minefield of Australian Uranium', *Journal of Australian Studies*, vol. 37, no. 4, 2013, pp. 442–43.
83 Gerster, 'Down the Yellowcake Road', p. 448.
84 'Uranium', *Morning Bulletin* (Rockhampton), 15 October 1953, p. 4.
85 Gerster, 'Down the Yellowcake Road', p. 438.
86 Cawte, *Atomic Australia*, p. 169.

Chapter 10 Marie Celeste

Epigraphs: Noel Williams, Statements from Australian witnesses: T–Z, Royal Commission into British Nuclear Tests in Australia, NAA: A6450, 5, p. 5; JF Hill, Report of enquiry into souveniring of official stores from Emu Field, addressed to deputy director, Melbourne, Commonwealth Investigation Service, 25 April 1954, NAA: A1533, 1954/299; CA Stinson, acting secretary, Department of Supply, Confidential memorandum to director, Commonwealth Investigation Service, Attorney-General's Department, 14 January 1954, NAA: A1533, 1954/299.

1 JL Symonds, *A History of British Atomic Tests in Australia*, AGPS, Canberra, 1985, p. 220.
2 Lorna Arnold & Mark Smith, *Britain, Australia and the Bomb: The Nuclear Tests and Their Aftermath*, 2nd edn, Palgrave Macmillan, New York, 2006, p. 67.
3 Ivan Southwell, *Woomera*, Angus & Robertson, Sydney, 1962, p. 164.
4 Royal Commission into British Nuclear Tests in Australia, *Report* (JR McClelland, president), 2 vols, AGPS, Canberra, 1985, p. 229.
5 Howard Beale, 'Future Atomic Tests in Australia', Top-secret letter to Robert Menzies, 12 November 1953, NAA: A6456, R075/002.
6 Beale, 'Future Atomic Tests in Australia'.
7 'Less Remote Atom Site Possible', *Advertiser* (Adelaide), 27 November 1953, NAA: A6456, R121/100.
8 LC Lucas, director of construction, Letter to chief superintendent, LRWE, Salisbury, 22 October 1953, NAA: A6456, R115/002.
9 Lucas, Letter.
10 LC Lucas, Confidential outward teletype message to Group Captain AG Pither, 30 October 1953, NAA: A6456, R115/002.
11 Barry Hodges & Graham Munsell, B Squadron 1st Armoured Regiment (retired), 'The Long Journey', Unpublished profile of the atomic tank, Gol Gol & Mirrabooka, 2019.
12 CF Bareford, chief superintendent, Restricted memorandum to controller, Research Trials and Services, 9 December 1953, NAA: A6456, R115/003.
13 Bareford, Restricted memorandum.
14 Bareford, Restricted memorandum.
15 Bareford, Restricted memorandum.
16 Stinson, Confidential memorandum.
17 Stinson, Confidential memorandum.
18 AG Pither, group captain, Confidential letter to superintendent, LRWE, Woomera, 8 December 1953, NAA: A6456, R115/002.
19 RC Simpson, manager, Department of Supply, Memorandum to controller, Weapons Research Establishment, Salisbury, 31 May 1955, NAA: A6456, R115/002.
20 Simpson, Memorandum.
21 Australian House of Representatives, *Debates*, 25 November 1953, p. 475 (Howard Beale).
22 Simpson, Memorandum.
23 Hill, Report.
24 Hill, Report.
25 Hill, Report.
26 Howard Beale, minister for supply, Speech to House of Representatives, 24 November 1953, Commonwealth of Australia, Hansard, p. 403.

27 Hill, Report.
28 Hill, Report.
29 Hill, Report.
30 Hill, Report.
31 Hill, Report.
32 Hill, Report.
33 Hill, Report.
34 Hill, Report.
35 Hill, Report.
36 Hill, Report.
37 Hill, Report.
38 DH Black, head, UK Ministry of Supply staff Australia, Personal and secret letter to Captain FB Lloyd, Ministry of Supply, 10 December 1953, Hill, Report, appendix 8.
39 Captain FB Lloyd, Ministry of Supply, Personal and confidential letter to DH Black, head, UK Ministry of Supply staff Australia, 22 December 1953, Hill, Report, appendix 8.
40 '"Souveniring" at Atom Test Site', *Advertiser* (Adelaide), 25 November 1953, NAA: A6456, R121/100.
41 E Hattam, deputy superintending peace officer, Commonwealth Investigation Service, Canberra, Letter to superintending peace officer, 23 September 1953, NAA: A6456, R121/100.
42 E Hattam, deputy superintending peace officer, Commonwealth Investigation Service, Canberra, Letter to EA Hayes, Peace Officer Guard, Woomera Detachment, 6 October 1953, NAA: A6456, R121/100.
43 Hattam, Letter to Hayes.
44 Hattam, Letter to Hayes.
45 E Hattam, Memo to superintending peace officer, Peace Officer Guard, South Australia, Woomera Detachment, LRWER: Security arrangements and protection at X.200 site, Project Emu Field (atomic blast area), 9 November 1953, NAA: A6456, R121/100.
46 Williams, Statements from Australian witnesses, p. 3.
47 CS Morrison, X200 project security officer, Secret memorandum to director, X200 Force, 21 October 1953, NAA: A6456, R121/100.
48 E Hattam, deputy superintending peace officer, Commonwealth Investigation Service, Canberra, Letter to superintending peace officer, 7 September 1953, NAA: A6456, R121/100.
49 CS Morrison, X200 project security officer, Post orders – Peace Officer Guard – X200 site, 6 November 1953, NAA: A6456, R121/100.
50 Thomas Murray, Transcript of proceedings, Royal Commission into British Nuclear Tests in Australia, NAA: A6448, 6, p. 2357.
51 Morrison, Post orders.
52 Morrison, Post orders.
53 Noel Williams, Transcript of proceedings, Royal Commission into British Nuclear Tests in Australia, NAA: A6448, 8, p. 4008.
54 Williams, Transcript of proceedings, p. 4008.
55 Williams, Statements from Australian witnesses, p. 3.
56 Williams, Statements from Australian witnesses, p. 5.
57 Williams, Statements from Australian witnesses, p. 5.

58   Williams, Transcript of proceedings, p. 4010.
59   Murray, Transcript of proceedings, p. 2357.
60   EA Hayes, officer in charge, Secret note to deputy superintending peace officer, 7 April 1954, NAA: A6456, R121/100.
61   Thomas Murray, Statements from Australian witnesses: K–O, Royal Commission into British Nuclear Tests in Australia, NAA: A6450, 3, p. 3.
62   Murray, Statements from Australian witnesses.
63   FJ McKenna, acting secretary, Prime Minister's Department, Top-secret letter to Ben Cockram, office of the High Commissioner for the United Kingdom, 22 April 1953, NAA: A6456, R124/018.
64   McKenna, Top-secret letter.
65   FA O'Connor, secretary, Department of Supply, Secret letter to assistant secretary, Department of the Treasury (Defence Division), 23 February 1954, NAA: A6456, R124/018.
66   O'Connor, Secret letter.
67   'Work on Guided Missiles at Woomera', *Nature*, vol. 173, no. 4385, 14 November 1953, p. 893.

Chapter 11 The forgetting of Emu
Epigraphs: Atomic Weapons Research Establishment, *Operation Totem*, film, Aldermaston, 1953, available at International War Museums, <www.iwm.org.uk/collections/item/object/1060036296>, accessed 2 February 2021; Karina Lester, quoted in Dimity Hawkins & Matthew Bolton, *Addressing Humanitarian and Environmental Harm from Nuclear Weapons: Monte Bello, Emu Field and Maralinga Test Sites*, International Disarmament Institute and Helene & Grant Wilson Center for Social Entrepreneurship, New York, 2018, p. 1.
1   Ministry of Supply meeting, Top-secret note, Ref. 330/10/88/1, St Giles Court, London, 28 February 1953, AB 16/1743, NAUK.
2   Nic Maclellan, *Grappling with the Bomb: Britain's Pacific H-Bomb Tests*, Australian National University Press, Canberra, 2017, p. 105.
3   Graham Farmelo, *Churchill's Bomb: A Hidden History of Science, War and Politics*, Faber & Faber, London, 2013, p. 442.
4   Farmelo, *Churchill's Bomb*, p. 442.
5   Farmelo, *Churchill's Bomb*, p. 442.
6   Farmelo, *Churchill's Bomb*, p. 443.
7   William Penney, Secret letter to JES Stevens, chairman, Totem Panel, 2 November 1953, NAA: A6456, R124/018.
8   Penney, Secret letter.
9   Penney, Secret letter.
10   Penney, Secret letter.
11   Penney, Secret letter.
12   Penney, Secret letter.
13   Penney, Secret letter.
14   Expert Committee on the Review of Data on Atmospheric Fallout Arising from British Nuclear Tests in Australia, *Report of the Expert Committee on the Review of Data on Atmospheric Fallout Arising from British Nuclear Tests in Australia* (CB Kerr, chairman; Kerr report), AGPS, Canberra, 31 May 1984, p. 33, digitised copy at International Atomic Energy Agency, <inis.iaea.org/

collection/NCLCollectionStore/_Public/17/019/17019465.pdf>, accessed 26 July 2021.

15 JR Moroney & KN Wise, ARL, 'Public Health Impact of Fallout from British Nuclear Weapons Tests in Australia, 1952–1957', Submission to the Royal Commission into British Nuclear Tests in Australia, July 1985, Titterton archive, box 101, series 22, folder 22/41, AAS.
16 JR Moroney & KN Wise, 'Impact on Public Health of Long-Range Fallout from Nuclear Weapons Tests in Australia, 1952–1957', Draft statement for Royal Commission, prepared for viewing by Ernest Titterton, 26 September 1984, Titterton archive, box 101, series 22, folder 22/41, AAS.
17 Ernest Titterton & Alan Butement, Letter to the editor, 19 November 1985, Titterton archive, box 100, series 22, AAS.
18 William J Gibbs, Letter to Ernest Titterton, 20 November 1985, Titterton archive, box 101, series 22, AAS.
19 Gibbs, Letter.
20 Gibbs, Letter.
21 DJ Stevens, Letter to Ernest Titterton, 6 November 1985, Titterton archive, box 101, series 22, AAS.
22 Titterton & Butement, 'Safety Levels at N-Tests', letter to the editor, *Canberra Times*, 30 November 1985, p. 16.
23 Ernest Titterton, Criticisms of the McClelland Royal Commission, undated, Titterton archive, box 98, folder 22/3, AAS.
24 KE Thompson, Department of Science and the Environment, Environment Division, Briefing for the minister for science on a media release on the Maralinga and Emu test sites, AIRAC report on residual radioactivity at Emu Field, 20 March 1980, NAA: A6456, R057/003.
25 RJ Walsh, chairman, AIRAC, Letter to JJ Webster, minister for science and the environment, 19 October 1979, NAA: A6456, R057/003.
26 Walsh, Letter.
27 Malcolm Fraser, prime minister, Letter to David Tonkin, premier of South Australia, 12 March 1980, NAA: A6456, R057/003.
28 Elizabeth Tynan, *Atomic Thunder: The Maralinga Story*, NewSouth, Sydney, 2016, p. 270.
29 DS Maclagan, MB Cooper & JC Duggleby, 'Residual Radioactive Contamination of Emu Plain from Nuclear Weapons Tests Conducted in 1953', Draft ARL report, August 1979, NAA: A6456, R057/003, Abstract, unpaginated.
30 Maclagan, Cooper & Duggleby, 'Residual Radioactive Contamination', p. 2.
31 Maclagan, Cooper & Duggleby, 'Residual Radioactive Contamination', p. 3.
32 Maclagan, Cooper & Duggleby, 'Residual Radioactive Contamination', pp. 10–11.
33 DP Child & MAC Hotchkis, 'Plutonium and Uranium Contamination in Soils from Former Nuclear Weapon Test Sites in Australia', *Nuclear Instruments and Methods in Physics Research B*, vol. 294, 2013, p. 643.
34 Child & Hotchkis, 'Plutonium and Uranium Contamination', p. 643.
35 Child & Hotchkis, 'Plutonium and Uranium Contamination', pp. 643–44.
36 Child & Hotchkis, 'Plutonium and Uranium Contamination', p. 646.
37 Child & Hotchkis, 'Plutonium and Uranium Contamination', p. 646.
38 Department of Education, Science and Training, *Rehabilitation of Former*

NOTES TO PAGES 270-278

Nuclear Tests Sites at Emu and Maralinga (Australia) 2003 (MARTAC report), Commonwealth of Australia, Canberra, 2002, p. 168.
39   Wayne Reynolds, 'Rethinking the Joint Project: Australia's bid for nuclear weapons, 1945–1960', Historical Journal, 41, 3, 1998, p. 858.
40   'Atom Power in 1953', *Mail* (Adelaide), 27 December 1952, p. 3.
41   'The Atomic Blast', *West Australian*, 16 October 1953, p. 2.
42   Peter McClellan, 'Who is telling the truth? Psychology, common sense and the law', *Australian Law Journal*, 2006, vol. 80 (10), p.655–666, p. 656.
43   Nuclear Weapon Archive, 'Operation Totem: 1953', in 'Britain's Nuclear Weapons: British Nuclear Testing', Nuclear Weapon Archive, last changed 23 August 2007, <nuclearweaponarchive.org/Uk/UKTesting.html>, accessed 28 January 2021.
44   Note from United Kingdom High Commission to Australia, Woomera – development trials, 4 June 1959, Athol Townley, minister for defence, Cabinet submission no. 466, Nuclear mechanisms – trials at Woomera, 6 November 1959, NAA: A5818, VOLUME 10/AGENDUM 466, annex A.
45   John McEwen, deputy prime minister, & Athol Townley, defence minister, Top-secret memo to Robert Menzies, prime minister, 13 June 1959, Townley, Cabinet submission, annex B.
46   McEwen & Townley, Top-secret memo.
47   Townley, Cabinet submission.
48   Cabinet minute, Decision no. 550, Townley, Cabinet submission.
49   Moroney & Wise, 'Impact on Public Health of Long-Range Fallout from Nuclear Weapons Tests in Australia, 1952–1957', p. iii.
50   Royal Commission into British Nuclear Tests in Australia, *Report* (JR McClelland, president), 2 vols, AGPS, Canberra, 1985, p. 229.
51   Tynan, Atomic Thunder, p. 2.
52   Noah Pearce, 'Final Report on Residual Radioactive Contamination of the Maralinga Range and the Emu Site', AWRE report no. 01–16/68 (Pearce report), UK Atomic Energy Authority, January 1968.
53   Pearce, 'Final Report', p. 5.
54   Tynan, Atomic Thunder, p. 233.
55   Pearce, 'Final Report', p. 5.
56   Pearce, 'Final Report', p. 6.
57   DEST, *Rehabilitation*, p. 75.
58   DEST, *Rehabilitation*, p. 76.
59   DEST, *Rehabilitation*, p. v.
60   DEST, *Rehabilitation*, p. 91.
61   DEST, *Rehabilitation*, p. 166.
62   DEST, *Rehabilitation*, p. 83.
63   DEST, *Rehabilitation*, p. 167.
64   DEST, *Rehabilitation*, p. 167.
65   DEST, *Rehabilitation*, p. 384.
66   Oak Valley Rangers, 'What, Who, How, Where: A Brief Summary of the OV 2020/21 Oak Valley Ranger Work Program', brochure, Oak Valley, 2020, p. 5.
67   Oak Valley Rangers, 'What, Who, How, Where', p. 1.

BIBLIOGRAPHY

A note about sources

Most of my primary sources in Australia have come from the National Archives of Australia (NAA) and the Australian Academy of Science (AAS); and in the United Kingdom, the National Archives of the United Kingdom in Kew (NAUK), the Portal Collection at Christ Church Library, Oxford (Portal Collection), the Lindemann/Cherwell papers at Nuffield College Library, Oxford (Lindemann/Cherwell papers) and the Churchill Archives Centre at Cambridge University (Churchill Archives). I have been fortunate to have access to other primary material as well. The former *Canberra Times* journalist Paul Malone received numerous files from government departments, notably the Department of Defence and the Department of Supply, under an accelerated document release program in the early 1980s. Paul generously gave me access to these files, which do not have NAA numbers. The same files may well have found their way to the NAA, but in some cases I have used the unnumbered files provided by Paul. I note these particular files in the relevant parts of the bibliographical details using the description 'Malone files'.

Media reports – Author unnamed (chronological order)
'Atom Power in 1953', *Mail* (Adelaide), 27 December 1952, p. 3.
'Britain "a factor in A-warfare"', *Argus*, 9 January 1953, Department of Defence. Atomic Energy – General NAA: A5954, 2163/1.
'A British Nevada', *Advertiser* (Adelaide), 26 June 1953, p. 2.
'New A-Bomb for Woomera; *Sunday Telegraph (Sydney)*, 21 July 1953, Atomic test at Woomera, NAA: A5954, 2167/1.
Herald Special Service, 'A-Bombs Flown Here', *Herald* (Melbourne), 22 July 1953, Atomic test at Woomera, NAA: A5954, 2167/1.
'Woomera's Big Role', *Herald* (Melbourne), 3 August 1953, Atomic test at Woomera, NAA: A5954, 2167/1.
'Woomera Atom Test in October: Sir William Penney Coming Again', *Age*, 3 August 1953, Atomic test at Woomera, NAA: A5954, 2167/1.
'America Can Watch This – "If...": Secret Atom Missiles Fly Here for October Test', *Argus*, 3 August 1953, Atomic test at Woomera, NAA: A5954, 2167/1.
'Talks on Atomic Weapons: Ministers Meet in Canberra', *Daily Telegraph*, 31 August 1953, Atomic test at Woomera, NAA: A5954, 2167/1.
'Woomera Trials in October', *Herald* (Melbourne), 4 September 1953, Atomic test at Woomera, NAA: A5954, 2167/1.
'US Reps at Atom Tests', *Sunday Sun & Guardian* (Sydney), 6 September 1953, Atomic test at Woomera, NAA: A5954, 2167/1.
'No Danger from Bomb Test: P.M. Assures of Safety', *Age*, 9 September 1953, Atomic test at Woomera, NAA: A5954, 2167/1.
'Co-operation at Woomera', *Sydney Morning Herald*, 15 September 1953, Atomic test at Woomera, NAA: A5954, 2167/1.
'His Responsibility!', *Daily Mirror (Sydney)*, 2 October 1953, p. 2.
'Atom Bombs in Our Arid Lands', *Sunday Herald*, 4 October 1953, p. 2.
'The Price of Uranium', Plain English editorial, *Bulletin*, vol. 74, no. 3843, p. 6.
'Atomic Bombs, Uranium and Australia's Future', *Sydney Morning Herald*, 8 October 1953, p. 2.
'Our Future Lies in this Asset', editorial, *Argus*, 8 October 1953, p. 2.
Staff correspondent, 'Public Safety and the Woomera Tests', *Sydney Morning Herald*, 9 October 1953, p. 2.
Staff correspondent, 'Long-range Risks in Atomic Tests?', *Sydney Morning Herald*, 10 October 1953, p. 2.
'Uranium', *Morning Bulletin* (Rockhampton), 15 October 1953, p. 4.
'A Flash and a Roar', *Advertiser* (Adelaide), 16 October 1953, p. 2.
A 'Herald' Special Reporter Who Watched the Atomic Explosion on the Woomera Range from 15 Miles Away, 'Atom Explosion Success: Dawn Blast at Woomera', *Sydney Morning Herald*, 16 October 1953, p. 1.
Special Correspondent in London, 'How an Atomic Bomb Explodes', *Sydney Morning Herald*, 16 October 1953, p. 2.
'Months of Hard, Lonely Work Paved Way for Atomic Explosion', *Sydney Morning Herald*, 16 October 1953, p. 4.
'The Atomic Blast', *West Australian*, 16 October 1953, p. 2.
'Second A-Blast Successful: Trial Series Ends', *Sydney Morning Herald*, 28 October 1953, p. 1.
'Other 'Minor' Tests', *Sydney Morning Herald*, 28 October 1953, p. 1.
'Work on Guided Missiles at Woomera', *Nature*, vol. 173, no. 4385, 14 November 1953, p. 893.

'"Souveniring" at Atom Test Site', *Advertiser* (Adelaide), 25 November 1953, NAA: A6456, R121/100.
'Less Remote Atom Site Possible', *Advertiser* (Adelaide), 27 November 1953, NAA: A6456, R121/100.
'Man Who Led Fighting Sixth is Dead', *Age*, 21 May 1969, included in Personal correspondence – Australia. Sir JES Stevens, secretary, Department of Supply, NAA: A5954, 70/11.
'Crews "Over-Exposed" and "Under-Protected" in A-Test', *Canberra Times*, 10 October 1984, p. 12.
'Atomic Tests "Threatened No-One"', *Canberra Times*, 15 May 1985, p. 3.

Media reports – Author named
Cameron, James, 'When the Desert Skies Caught Fire', *Age*, 16 October 1953, p. 2.
Dvoretsky, Harold, 'Choice of a Weapon: Woomera Test Experts in Big Decision', *Sunday Telegraph (Sydney)*, 16 August 1953, page unknown, NAA: A5954, 2167/1.
Leigh, David, & Lashmar, Paul, 'Forgotten victims of the "Rolling Black Mist", originally published in *Observer*, edited and republished in *Bulletin*, vol. 103, no. 5360 12 April 1984, pp. 97–100.
Malone, Paul, & Conkey, Howard, 'Impact on Aborigines: The Loud Bang and the Black Cloud Linger', *Canberra Times*, 29 September 1984, p. 15.
Monson, Ronald, 'Security Was Tight on Project X200', *Daily Telegraph (Sydney)*, 24 October 1953, p. 21.
Smith, Trevor, 'A Test Here in October: U.K. Minister for Woomera', *Herald* (Melbourne), 25 July 1953, Atomic test at Woomera, NAA: A5954, 2167/1.

Primary sources
Aboriginal Health Unit, South Australian Health Commission, *A Survey of Diseases That May Be Related to Radiation Among Pitjantjatjara on Remote Reserves*, 18 February 1981.
Adams, CA, AWRE, Confidential letter to Ernest Titterton, 14 December 1953, Titterton archive, box 100, series 22, AAS.
Adams, CA, AWRE, Document showing Totem staff, [1953], AB 16/1743, NAUK.
Adams, CA, AWRE, Letter to Ernest Titterton, 11 September 1953, Titterton archive, box 100, series 22, AAS.
AIRAC Report on the Emu Atomic Test sites, Cover letter addressed to the Minister for Science, 12 March 1980, Department of Environment, Housing and Community Development, Atomic Energy Commission – Report on residual radioactivity at Emu Field, NAA: A6456, R057/003.
'Atomic Explosion: Message from Sir Winston Churchill', Press notice no. 132/1953, 16 October 1953, NAA: A6456, R124/018.
Atomic Test – Montebello. Notes of meeting of Dr Solandt, chairman Canadian Defence Research Board, and Professor Martin, defence scientific adviser, with Defence Committee, 9 October 1952, NAA: A5954, 7/11.
Atomic warfare resources, Report no. 32/1949, Defence Research and Development Policy Committee, 7 December 1949, NAA: A5954, 2330/2.
Attlee, Clement, Letter to Winston Churchill, 3 December 1950, CHUR 2/28, Churchill Archives.
Austin, J, AWRE, Operation Totem team report, Team RH5 – decontamination, The prevention and removal of radio-active contamination, part VI, Decontamination

of aircraft and health control at Woomera and Amberley, 6 September 1954, NAA: A6454, Z100.

Bailey, David, Handwritten note to officer in charge, Defence Records, 18 April 1997, NAA: A12372, A21830, BAILEY, David Richard, Service record.

Bareford, CF, chief superintendent, Restricted memorandum to controller, Research Trials and Services, 9 December 1953, NAA: A6456, R115/003.

Barnard, Geoffrey, Deputy Chief of the Naval Staff, top-secret letter to Air Marshal Sir Ronald Ivelaw-Chapman, Deputy Chief of the Air Staff, 5 August 1953, AIR 2/13778, NAUK.

Barnes, DE, GC Dale, RA Siddons, & ER Woodcock, High Explosives Research report no. A32, 'Airborne Contamination at the Totem Trial', Materials and Physical Research Division, High Explosives Research, UK Ministry of Supply, May 1953, NAA: A6454, ZB3.

Beadell, Len, Transcript of proceedings, Royal Commission into British Nuclear Tests in Australia, NAA: A6448, 7.

Beale, Howard, Minister for Supply, Top-secret letter to Robert Menzies, prime minister, 28 May 1953. NAA: A4940 C1730 PART 1.

Beale, Howard, 'Future Atomic Tests in Australia', Top-secret letter to Robert Menzies, 12 November 1953, NAA: A6456, R075/002.

Beale, Howard, Minister for Supply, Speech to the House of Representatives, 24 November 1953, Commonwealth of Australia Hansard, p. 403.

Beale, Howard, Minister for Supply, Speech to the House of Representatives, 25 November 1953, Commonwealth of Australia Hansard, p. 475.

Beale, Howard, Minister for Supply, 'Atomic Tests in Australia', Top-secret Cabinet briefing document, Submission no. 73, Canberra, 11 August 1954, Malone files.

Beale, Howard, Minister for Supply, Top-secret letter to Percy Spender, Australian ambassador to Washington, 22 October 1954, NAA: A6456, R065/011.

Black, DH, head of UK Ministry of Supply Staff Australia, Personal and secret letter to Captain FB Lloyd, Ministry of Supply, 10 December 1953, Appendix 8, JF Hill, Report of enquiry into souveniring of official stores from Emu Field, addressed to deputy director, Melbourne, Commonwealth Investigation Service, 25 April 1954, NAA: A1533, 1954/299.

Brinkley, TA, 'Atomic Bomb Test Series at X200 Site October 1953: Recollections of Totem I', Titterton archive, box 99, series 22, AAS.

Brown, AS, secretary, Prime Minister's Department, Top-secret personal note to Ben Cockram, UK High Commission, 18 March 1953, NAA: A6456, R124/018.

Brown, AS, secretary, Prime Minister's Department, Secret note to Prime Minister Robert Menzies, 23 September 1953, NAA: A6456, R065/003.

Brown, AS, secretary, Prime Minister's Department, Secret note to Prime Minister Robert Menzies, 28 September 1953, NAA: A6456, R065/003.

Brown, AS, secretary, Prime Minister's Department, Confidential letter to JES Stevens, 29 September 1953, NAA: A6456, R065/003.

Brown, AS, secretary, Prime Minister's Department, Top-secret note to Robert Menzies, 6 October 1953, NAA: A4949, C2047.

Brown, AS, secretary, Prime Minister's Department, 9 March 1954, NAA: A6706, 50.

Buchanan, AE, '"D" Notice No. 10', Defence, Press and Broadcasting Committee, NAA: A1209, 1957/5486.

Burrage, Restricted outward teletype message to Garden, 29 October 1953, NAA: A6456, R115/002.

Cabinet Committee on Uranium, Programme for atomic development, Agendum no. UC/8, May 1953, NAA: A4940 C1730 Part 1.
Cabinet Committee on Uranium, Top-secret minutes, Attached draft notes in response to meeting with Lord Cherwell, 6 October 1953, NAA: A4949, C2047.
Cabinet Minute, Decision no. 550, Submission no. 466, Nuclear Mechanisms – Trials at Woomera, 19 November 1959, NAA: A5818, vol. 10/agendum 466.
Carruthers, JA, RJ Wilson, PC East, & EJ Ellis, Ellis, AWRE report no. T 10/55, Thermal measurements at Operation Totem, ES 5/125, NAUK.
Carter, Leo, Top-secret memorandum to chairman, Totem Panel, 31 March 1953, NAA: A6456, R215/016.
Carver, John, Letter to Ernest Titterton from Salisbury, South Australia, undated, Titterton archive, box 100, series 22, AAS.
Chadwick, Sir James, Notes on draft memorandum by Portal, 5 August 1946, Lord Portal archive, Atomic energy folder B, UK Military Atomic Energy Chronology 1945–1969 (including V bomber force), Portal Collection.
Chadwick, Sir James handwritten note to Lord Portal, 24 October 1966, in response to a letter for Chadwick's 75th birthday, Lord Portal archive, Atomic Energy folder B2, UK Military Atomic Energy Chronology 1945–1969 (including V bomber force), Portal Collection.
Chadwick, James, Top-secret letter to Lord Portal, 'Private and Personal', 8 April 1946, Lord Portal archive, Atomic Energy folder B2, UK Military Atomic Energy Chronology 1945–1969 (including V bomber force), Portal Collection.
Chadwick, James, Note to JPW Ehrman, Cabinet Office London, 23 December 1952. CHAD IV 12/5 Chadwick papers, Churchill Archives.
Churchill, Winston, letter to Harry S Truman, 12 February 1951, CHUR 2/28, Churchill Archives.
Churchill, Winston, Top-secret cabinet document 14 December 1954: 'Notes on Tube Alloys, 1954'. DSND 14/4, Churchill Archives.
Clark, Allen, Statements from Australian Witnesses A–D, Royal Commission into British Nuclear Tests in Australia, NAA: A6450, 1.
Clarke QC, John, Air Marshal Doug Riding & Dr David Rosalk, 'British Atomic Tests', Chapter 16 in *Report of the Review of Veterans' Entitlements*, Commonwealth of Australia, Canberra, 6 January 2003.
Cockram, Ben, Top-secret letter to Neil Pritchard, 19 January 1953, AB 16/1743, NAUK.
Cockram, Ben, Top-secret letter to Robert Menzies, 3 March 1953, NAA: A6456, R124/018.
Cockram, Ben, Confidential letter to Brown, 9 November 1953, DEFE 16/819, NAUK.
Cockram, Ben, Top-secret letter to Neil Pritchard, 25 November 1955, FCO 1/6, NAUK.
Colquhoun, Group Captain DW, officer commanding, Secret letter to director, Air Force Intelligence, Melbourne, 19 November 1953, NAA: A13064, DPIE 92/001390.
Colquhoun, David, Transcript of proceedings, Royal Commission into British Nuclear Tests in Australia, NAA: A6448, 15.
Commonwealth Department of Veterans' Affairs, Letter to Morris Newman, 13 June 1996, provided by Michele Newman.
Commonwealth Relations Office, Telegram to UK High Commissioner in New Zealand, [1953], AB 16/1743, NAUK.

BIBLIOGRAPHY 323

Cooke, JA, on behalf of acting regional manager, Department of Aboriginal Affairs South Australian Region, 'Aboriginals and Atomic Tests', Letter to secretary, Department of Aboriginal Affairs, Canberra, 4 September 1980, Malone files.

Cosgrove, Roy, Statements from Australian Witnesses: A–D, Royal Commission into British Nuclear Tests in Australia, NAA: A6450, 1.

Crofts, JM, Australian Military Forces Northern Command, Letter to Ernest Titterton, 31 October 1953, Titterton archive, box 100, series 22, AAS.

Crofts, JM, Reports on radiation survey on the north east coast, 31 October 1953, Titterton archive, box 100, series 22, AAS.

CRTS, restricted inward teletype message to Chief Superintendent LRWE, 5 November 1953, NAA: A6456, R115/002.

Cunliffe-Lister, Philip, secretary of state for Commonwealth relations, Private note to Robert Menzies, 1 July 1953, NAA: A4949, C2047.

Daley, EA, air vice marshal, director general, Medical Services, RAAF, Radiological Health During Operation 'HURRICANE' Monte Bello Is. October/November 1952 and Operation 'TOTEM', Emu Claypan, October/November 1953, Department of Air secret minute paper, 3 December 1953, NAA: A6456, R021/001 PART 79.

Daley, EA, air vice marshal, director general, Medical Services, RAAF, Secret minute, 25 May 1954, Department of Air – Medical implications in the RAAF, NAA: A6456, R024/001.

Defence Committee Agendum no. 3/1950, Secret notes, Atomic Warfare Resources, NAA: A5954, 2330/2.

Department of Education, Science and Training, *Rehabilitation of Former Nuclear Tests Sites at Emu and Maralinga (Australia) 2003* (MARTAC report), Commonwealth of Australia, Canberra, 2002.

Department of Environment, Housing and Community Development, Atomic Energy Commission, Report on Residual Radioactivity at Emu Field, February 1980, NAA: A6456, R057/003.

Department of Supply, Souveniring of official stores at Emu Field, 18 Dec 1953, NAA: A6456, R086/011.

Draft press notice, 19 February 1951, annotated by Winston Churchill, CHUR 2/28, Churchill Archives.

Draft press notice, 19 February 1951, annotated by Frederick Lindemann, CHUR 2/28, Churchill Archives.

Drake Seager, ER, Army Equipment Group, War Office, Report on Operation Totem, 14 December 1953, ES 1/834, NAUK.

Drake Seager, ER, Letter to CA Adams, 16 December 1953, ES 1/834, NAUK.

Drake Seager, ER, group leader, AE Group, Report prepared for director, AWRE, 29 January 1954, ES 18/25, NAUK.

Drake Seager, ER, Statements from UK Witnesses: A–L, Royal Commission into British Nuclear Tests in Australia, NAA: A6449, 1.

Duties of Mabel Creek Detachment, Statement of duties, 11 June 1953, NAA: A6456, R121/100.

Eastern Area Detachment Woomera, Restricted teletype message to 82 Wing, 28 October 1953, NAA: A13064, DPIE 92/001390.

Eames, Geoffrey & Andrew Collett, Final submission on behalf of Aboriginal groups and individuals, Royal Commission into British Nuclear Tests in Australia, 16 September 1985, NAA: A6455, RC862.

Eames, Geoffrey, Counsel for Aboriginal people, Opening remarks at Marla Bore, 20 April 1985, Transcript of proceedings, Royal Commission into British Nuclear Tests in Australia, NAA: A6448, 13.

Eames, Geoffrey, Counsel for Aboriginal people, Opening remarks on the first day of hearings in London, Transcript of proceedings, Royal Commission into British Nuclear Tests in Australia, NAA: A6448, 9.

Earl Marshal's Office, Belgrave Square, London, Letter to Elmhirst, 16 January 1953. ELMT 4/2 Elmhirst files, Churchill Archives.

Earner, Mark, Transcript of proceedings, Royal Commission into British Nuclear Tests in Australia, NAA: A6448, 2.

Editor, editorial in Bulldust (Emu Field), 8 August 1953, from the collection of the Curator, British Nuclear Test Veterans' Association, Wesley Perriman.

Edwards, Lancelot, Statements from Australian Witnesses: E–J, Royal Commission into British Nuclear Tests in Australia, Submission to the Royal Commission, NAA: A6450, 2.

Egan, L G, assistant crown solicitor, 'Woomera Rocket Range', Memorandum to assistant secretary, R&D, Department of Supply, 15 April 1953, NAA: A6456, R215/016.

Elmhirst, Thomas, Top-secret note to FC How, UK Ministry of Supply, 12 March 1953, AB 16/1743, NAUK.

Elmhirst, Thomas, confidential note to JM Wilson, Assistant Secretary, Establishments Division. 7 October 1953. ELMT 4/2 Elmhirst files, Churchill Archives.

Fadden, AW, Acting Prime Minister, Draft memorandum to all ministers, 22 June 1953, NAA: A6456, R124/018.

Fadden, AW, 'UK Atomic Test Woomera – D-notice to the Press', letter to Press Chiefs, 26 June 1953, in JL Symonds, A History of British Atomic Tests in Australia, AGPS, Canberra, 1985.

Fraser, Malcolm, prime minister, Letter to David Tonkin, premier of South Australia, 12 March 1980, NAA: A6456, R057/003.

Garrett, Lieutenant-General Sir Alwyn, chief of the general staff, Letter to Major General Eric Woodward, adjutant-general, Australian Army, 24 June 1953, NAA: A6456, R030/174.

Gates, BG, Department of Defence, Inward teletype message to FP O'Grady, assistant to CF Bareford, Chief Superintendent LWRE, 6 November 1953, NAA: A6456, R115/002.

GEN.163 minutes (handwritten), meeting of 8 January 1947, No. 10 Downing Street, London, Lord Portal archive, Atomic Energy folder B, UK Military Atomic Energy Chronology 1945–1969 (including V bomber force), Portal Collection.

Geschke, Charles, Statements from Australian witnesses: E–J, Royal Commission into British Nuclear Tests in Australia, NAA: A6450, 2.

Gibbs, William J, Letter to Ernest Titterton, 20 November 1985, Titterton archive, box 101, series 22, AAS.

Giles, Ellen, Statements from Australian Witnesses: E–J, Royal Commission into British Nuclear Tests in Australia, NAA: A6450, 2.

Gilfedder, D, Australian Atomic Energy Commission, Briefing notes for the Prime Minister's visit to UK and USA, 1955, Uranium production in Australia, 31 December 1954, NAA: A6456, R065/011.

BIBLIOGRAPHY   325

Gilfedder, D, Australian Atomic Energy Commission, Briefing notes for the Prime Minister's visit to UK and USA, 1955, 'International Agency', 31 December 1954, NAA: A6456, R065/011.

Gilfedder, D, Australian Atomic Energy Commission, Briefing notes for the Prime Minister's visit to UK and USA, 1955, 'Exchange of Information – Australia and the United Kingdom', 31 December 1954, NAA: A6456, R065/011.

Hattam, E, deputy superintending peace officer, Commonwealth Investigation Service, Canberra, Letter to superintending peace officer, 28 May 1953, NAA: A6456, R121/100.

Hattam, E, deputy superintending peace officer, Commonwealth Investigation Service, Canberra, Letter to superintending peace officer, 23 September 1953, NAA: A6456, R121/100.

Hattam, E, deputy superintending peace officer, Commonwealth Investigation Service, Canberra, Letter to EA Hayes, Peace Officer Guard, Woomera Detachment, 6 October 1953, NAA: A6456, R121/100.

Hayes, EA, officer in charge, Secret note to deputy superintending peace officer, 7 April 1954, NAA: A6456, R121/100.

HEROD Committee, agenda for eighth meeting to be held on 20 July 1953, AIR 2/13778, NAUK.

HEROD Committee, Minutes of the 8th meeting held at Air Ministry, Whitehall, chaired by Sir Ronald Ivelaw-Chapman, 20 July 1953, AIR 2/13778, NAUK.

High Explosives Research, Notes of a meeting, 28 November 1952, DEFE 16/454, NAUK.

Hill, JF, Report of enquiry into souveniring of official stores from Emu Field, addressed to deputy director, Melbourne, Commonwealth Investigation Service, 25 April 1954, NAA: A1533, 1954/299.

How, FC, UK Ministry of Supply, Official Committee on Atomic Energy, Top-secret briefing document on Operation Totem, 2 February 1953, AB 16/1743, NAUK.

Indigenous people from Wallatinna, Transcript of proceedings, Royal Commission into British Nuclear Tests in Australia, NAA: A6448, 13.

Instructions to Captain Crofts for radiation survey of the north-east coast, Secret memorandum, undated and unsigned, Titterton archive, box 100, series 22, AAS.

Jeffery, KH, DCAS, Top-secret note to squadron leader Boyd, secretary of the Herod Committee, 15 October 1953, AIR 2/13778, NAUK.

Jones, RO, Air Marshal, chair, SALOME Committee, Top-secret minute to ACAS, 22 June 1953, AIR 2/13778, NAUK.

Kanytji & Pingkayi, Transcript of proceedings, Royal Commission into British Nuclear Tests in Australia, NAA: A6448, 13, pp. 7189–7190.

Kerr, Charles, chair, *Report of the Expert Committee on the Review of Data on Atmospheric Fallout Arising from British Nuclear Tests in Australia*, AGPS, Canberra, 31 May 1984.

Lander, Almerta, Transcript of proceedings, Royal Commission into British Nuclear Tests in Australia, NAA: A6448, 13, p. 7098.

Lennon, Lallie, Transcript of proceedings, Royal Commission into British Nuclear Tests in Australia, NAA: A6448, 13, p. 7147.

Lester, Yami, Transcript of proceedings, Royal Commission into British Nuclear Tests in Australia, NAA: A6448, 13, pp. 7080a–7540.

Lindemann, Frederick, Note to Churchill about the British position on the atomic bomb, 19 January 1950. Lindemann/Cherwell papers J97.

Lindemann, Frederick, Note to Churchill, 17 September 1952, Lindemann/Cherwell papers J122.
Lindemann, Frederick, Draft letter to Australian Prime Minister Robert Menzies, 11 December 1952, PREM 11/562, NAUK.
Lindemann, Frederick, ATOMIC ENERGY, Future U.K. Programme, Note to Churchill, 29 December 1952, Lindemann/Cherwell papers, J122.
Lindemann, Frederick, ATOMIC ENERGY, Future Anglo-American Relations, Secret note to Winston Churchill, 29 December 1952, PREM 11/290, NAUK.
Lindemann, Frederick (Lord Cherwell), Note to Thomas Elmhirst, 19 January 1953, ELMT 4/2, Elmhirst files, Churchill Archives.
Lindemann, Frederick (Lord Cherwell), Secret note to Mr FC How, UK Ministry of Supply, 6 February 1953, AB 16/1743, NAUK.
Lindemann, Frederick, Note to Churchill, 4 May 1953, Lindemann/Cherwell papers J138.
Lindemann, Frederick, Note to Churchill, 4 June 1953, Lindemann/Cherwell papers J138.
Lindemann, Frederick, Note to Churchill, 23 June 1953, Lindemann/Cherwell papers J138.
Lindemann, Frederick, Note to Churchill, 26 August 1953, Lindemann/Cherwell papers, J97.
Lindemann, Frederick, Note 1 to Churchill, 1 December 1953, Lindemann/Cherwell papers.
Lindemann, Frederick, Note 2 to Churchill, 1 December 1953, Lindemann/Cherwell papers.
Lindemann, Frederick, Draft reply to question from Edelman, 10 November 1953, Lindemann/Cherwell papers J138.
Lindemann, Frederick, Note to Churchill, 12 November 1953, Lindemann/Cherwell papers, J138.
Lindemann, Frederick, Letter to the editor, *The Times*, London, 21 May 1957, Lindemann/Cherwell papers E60.
Lloyd, Captain FB, Ministry of Supply, London, Top-secret letter to Neil Pritchard, Commonwealth Relations Office, 7 December 1953, FCO 1/6, NAUK.
Lloyd, FB, captain, Ministry of Supply, Personal and confidential letter to DH Black, head, UK Ministry of Supply Staff Australia, 22 December 1953, Appendix 8, JF Hill, Report of enquiry into souveniring of official stores from Emu Field, addressed to deputy director, Melbourne, Commonwealth Investigation Service, 25 April 1954, NAA: A1533, 1954/299.
Lord Portal archive, Atomic Energy folder B, UK Military Atomic Energy Chronology 1945–1969 (including V bomber force), Portal Collection.
Lucas, LC, director of construction, X200, Secret letter to the Secretary, Department of Supply, Melbourne, 5 March 1953, DEFE 16/679, NAUK.
Lucas, LC, Director of Construction, X200, Letter to chief superintendent, LRWE Salisbury, 22 October 1953, NAA: A6456, R115/002.
Lucas, LC, Director of Construction, X200, Confidential outward teletype message to Group Captain AG Pither, 30 October 1953, NAA: A6456, R115/002.
Lucas, LC, Director of Construction, X200, Secret cable to Cook, 20 July 1053, DEFE 16/679, NAUK.
MacDougall, WB, 'Detailed Survey of the Jangkuntjara Tribe – Their Traditional Tribal Country and Ceremonial Grounds', sent to superintendent, Woomera, 2 March 1953, Malone files.

Maclagan, DS, Cooper, MB & Duggleby, JC, 'Residual Radiation Contamination of Emu Plain from Nuclear Weapons Tests Conducted in 1953', Draft ARL report, August 1979, NAA: A6465, R057/003.

Martin, Leslie, University of Melbourne, Top-secret letter to Prime Minister Robert Menzies, 17 June 1953, NAA: A6456, R095/012.

McEnhill, J, UK Ministry of Supply, Operation Totem summary plan, part 1, NAA: A6454, ZB1.

McKenna, FJ, acting secretary, Prime Minister's Department, Top-secret letter to Ben Cockram, officer of the Higher Commissioner for the United Kingdom, 22 April 1953, NAA: A6456, R124/018.

McKnight, AD, Prime Minister's Department, Letter to George Davey, office of the High Commissioner for the United Kingdom, Canberra, 30 September 1952, NAA: A6456, R096/006.

McKnight, AD, Prime Minister's Department, Top-secret personal letter to Ben Cockram, UK High Commission to Australia, 23 January 1953, NAA: A6456, R124/018.

McKnight, AD, Prime Minister's Department, File note, 29 April 1953, NAA: A6456, R124/018.

Menzies, Robert, Personal letter to Winston Churchill, 9 November 1953, EG 1/35, NAUK.

Menzies, Robert, Top-secret personal letter to marquis of Salisbury, 17 March 1954, NAA: A6706, 50.

Messenger, Lt Col W de LM, AWRE report no. T78/54, Operation Totem: the effects of an atomic explosion on a Centurion tank, vol. 1, December 1956, ES 5/77, NAUK.

Ministry of Supply meeting, Top-secret note, Ref. 330/10/88/1, St Giles Court, London, 28 February 1953, AB 16/1743, NAUK.

Minutes of the seventh meeting of interdepartmental [Totem] panel, Top-secret, Melbourne, 24 June 1953, NAA: A6456, R095/012.

Morrison, CS, X200 project security officer, Secret memorandum to director, X200 Force, 21 October 1953, NAA: A6456, R121/100.

Morrison, CS, X200 project security officer, Post orders – Peace Officer Guard – X200 site Emu, NAA: A6456, R121/100.

Moroney, JR, & Wise, KN, 'Impact on Public Health of Long-Range Fallout from Nuclear Weapons Tests in Australia, 1952–1957', Draft statement for the Royal Commission, prepared for viewing by Ernest Titterton, 26 September 1984, Titterton archive, box 101, series 22, folder 22/41, AAS.

Moroney, JR, & Wise, KN, Australian Radiation Laboratory, 'Public Health Impact of Fallout from British Nuclear Weapons Tests in Australia, 1952–1957', Submission to the Royal Commission into British Nuclear Tests in Australia, July 1985, Titterton archive, box 101, series 22, folder 22/41, AAS.

Murray, Thomas, Statements from Australian witnesses: K–O, Royal Commission into British Nuclear Tests in Australia, NAA: A6450, 3.

Murray, Thomas, Transcript of proceedings, Royal Commission into British Nuclear Tests in Australia, NAA: A6448, 6.

Naggs, Rex, Transcript of proceedings, Royal Commission into British Nuclear Tests in Australia, NAA: A6448, 4.

Nettley, Richard, retired RAF, Statements from UK witnesses: M–Z, Royal Commission into British Nuclear Tests in Australia, NAA: A6449, 2.

Newman, Morris, Death certificate, Registration number 2000/60765, Queensland Government, 2000, provided by Michele Newman.
Newman, Morris, Statements from Australian witnesses: K–O, Royal Commission into British Nuclear Tests in Australia, NAA: A6450, 3.
Notes on an 'off the record' talk between Professor L Martin, Dr Butement and Dr WG Penney, 24 April 1952, NAA: A5954, 7/11.
O'Connor, FA, secretary, Department of Supply, Secret letter to assistant secretary, Department of the Treasury (Defence Division), 23 February 1954, NAA: A6456, R124/018.
Official guide for visitors to X200, 1953, NAA: A6456, R233/001.
United States Offices of the Joint Committee on Atomic Energy, Media release, 20 December 1954, NAA: A5462, 138/2/1.
Oliphant, Mark, Personal letter to Ernest Titterton, 6 October 1949, Titterton archive, box 1, MS 168, AAS.
Operation Totem – Emu Field 1953, Details of RAAF Participation in British Atomic Tests held in Australia 1952–1957, NAA: A6456, R021/001.
Operation 'Totem' consolidated report, RAAF, Headquarters Home Command, Penrith, 5 March 1954, NAA A6455, RC36, p. 13.
Pearce, Noah, 'Final Report on Residual Radioactive Contamination of the Maralinga Range and the Emu Site', AWRE report no. 01–16/68 (Pearce report), UK Atomic Energy Authority, January 1968. Full version tabled by Australian Senate, May 1984.
Penney, William, Top-secret letter to Lord Cherwell, undated, NAA: A6455, AB8.
Penney, William, Restricted de-briefing notice for UK staff, undated, Titterton archive, box 100, series 22, AAS.
Penney, William, Top-secret letter to Air Chief Marshal Sir Ralph Cochrane, 29 August 1952, AIR 2/13778, NAUK.
Penney, William, Personal note to Ernest Titterton, 22 May 1953, Titterton archive, box 100, series 22, AAS.
Penney, William, Secret letter to JES Stevens, chairman, Totem Panel, 2 November 1953, NAA: A6456, R124/018.
Penney, William, Top-secret letter to FC How, UK Ministry of Supply, 18 December 1952, AB 16/1743, NAUK.
Penney, William, Transcript of proceedings, Royal Commission into British Nuclear Tests in Australia, NAA: A6448, 9.
Penney, William, Transcript of proceedings, Royal Commission into British Nuclear Tests in Australia, NAA: A6448, 12.
Penney, William, Transcript of interview with Alfred Goldberg, United States Air Force Historical Division, 16 June 1963, AIR 2/13778, NAUK.
Pither, AG, group captain, Confidential letter to Superintendent, LRWE, Woomera, 8 December 1953, NAA: A6456, R115/002.
Press statement, 'Atomic Test Base Workers Held Coronation Parade', undated, NAA: A6456, R095/012.
Press statement, 'Australia's Part in Preparations for Atomic Test', 25 June 1953, NAA: A6456, R095/012.
Prime Minister's Department, Top-secret note, 13 March 1953, NAA: A6456, R124/018.
Prime Minister's Department, Top-secret note titled TOTEM, quoting letter from UK High Commission, 13 March 1953, NAA: A6456, R124/01.

BIBLIOGRAPHY   329

Pritchard, Neil, British deputy higher commissioner to Australia, Top-secret letter to Ben Cockram, 2 March 1953, AB 16/1743, NAUK.

Project X200 Progress Report no. 11 for week ended 6 June 1953, Top-secret note to chairman, Totem Panel, 10 June 1953, NAA: A6456, R096/002.

Quebec Agreement, reproduced in Margaret Gowing, Britain and Atomic Energy 1939–1945, London, Macmillan, 1964, pp. 439–40.

Redmond, TWJ, UK staff, Letter to Ernest Titterton, 11 September 1953, Titterton archive, box 100, series 22, AAS.

Roberts, DN, Air Commodore, director of operations, Top-secret note to DCAS, 15 January 1953, AIR 2/13778, NAUK.

Royal Australian Air Force Eastern Area Detachment Woomera, Operation order no. 1/53, RAAF operational information, Operation Totem, No. 6 (B) Squadron, file number 28/7/AIR, NAA: A11288, 28/7/AIR PART 1.

Royal Australian Air Force Operational Information, Operation Totem, no. 6 (B) Squadron, NAA: A11288, 28/7/AIR PART 1.

Royal Australian Air Force, No. 82 (B) Wing, Operation order no. 3/53, Operation 'Totem', NAA: A11288, 28/7/AIR PART 1.

Royal Australian Air Force, Assessment of conduct and trade proficiency for David Bailey, 9 September 1956, NAA: A12372, A21830, BAILEY, David Richard, Service record.

Royal Australian Air Force, Service conduct sheet, David Richard Bailey, 16 September 1956, NAA: A12372, A21830, BAILEY, David Richard, Service record.

Royal Commission into British Nuclear Tests in Australia, *Report: Conclusions and Recommendations* (JR McClelland, president, J Fitch and WJA Jonas, commissioners), AGPS, Canberra, 1985.

Royal Commission into British Nuclear Tests in Australia, Report (JR McClelland, president), 2 vols, AGPS, Canberra, 1985.

Sandys, Duncan, Letter to William Penney via the Ministry of Supply, 12 September 1953, NAA: A1533, 1954/299.

Sandys, Duncan, Minister of Supply, House of Commons Hansard, vol. 520, Monday 9 November 1953.

Sandys, Duncan, *Defence: outline of future policy* (Defence White Paper), Her Majesty's Stationery Office, London, March 1957, NAUK, CAB/129/86.

Secretary, Aborigines Protection Board, Adelaide, Memorandum to Walter MacDougall, native patrol officer, 26 August 1953. Malone files.

Secret minutes, Cabinet meeting of ministers, Cabinet Office, Great George Street, SW1, 12 November 1952, CAB 130/79/GEN417, NAUK.

Secret departmental briefing note, 'Discussions with Mr. Gordon Dean and Mr. Lewis Strauss, the Prime Minister and Mr. Dean', undated, NAA: A6456, R065/011.

Secret departmental briefing note, Notes furnished to the Australian Ambassador in Washington on the bilateral agreement between Australia and the U.S.A, undated, NAA: A6456, R065/011.

Secretary of State for Commonwealth Relations, Top-secret telegram to High Commissioner for the UK, Australia, 2 April 1955, NAUK, FCO/1/8.

Service department requirements for tests in future atomic bomb trials, Notes from meeting of DRP (AES)/M(53), 20 January 1953, AB 16/1743, NAUK.

Shedden, Frederick, secretary, Department of Defence, Top-secret letter to AD McKnight, Prime Minister's Department, 15 September 1952, NAA: A6456, R096/006.

Siddons, RA, Atomic Weapons Research Establishment Report no. E1/54, The rise of the cloud produced in an atomic explosion, Aldermaston, October 1954, p. 15, ES 3/6, NAUK.

Simpson, RC, manager, Department of Supply, Memorandum to controller, Weapons Research Establishment, Salisbury, 31 May 1955, NAA: A6456, R115/002.

S.O. Book 130, handwritten diary containing Totem data, ES 18/24, NAUK.

Stevens, DJ, Letter to Ernest Titterton, 6 November 1985, Titterton archive, box 101, series 22, AAS.

Stevens, JES, Top-secret letter to Ernest Titterton, 17 April 1953, NAA: A6456, R215/016.

Stevens, JES, Top-secret letter to Allan McKnight, 7 May 1953, NAA: A6456, R215/016.

Stevens, JES, Top-secret teleprinter message to Elmhirst, 15 May 1953, DEFE 16/523, NAUK.

Stevens, JES, chairman, Australian Atomic Energy Commission, Secret report prepared for the minister for external affairs, Notes on proposal for an international atomic energy pool as it affects Australia, 21 April 1954, NAA: A6456, R065/011.

Stevens, JES, Department of Supply, Melbourne, Secret teleprinter messages to Elmhirst, 29 July 1953, DEFE 16/679, NAUK.

Stinson, CA, acting secretary, Department of Supply, Confidential memorandum to director, Commonwealth Investigation Service, Attorney-General's Department, 14 January 1954, NAA: A1533, 1954/299.

Strauss, George, Chapter 12 Ministry of Supply – Atomic Bomb, from unpublished autobiography, STRS 1/2 Strauss papers, Churchill Archive Centre, University of Cambridge.

Thompson, KE, Department of Science and the Environment, Environment Division, Briefing for the minister for science on a media release on the Maralinga and Emu test sites, AIRAC report on residual radioactivity at Emu Field, 20 March 1980, NAA: A6456, R057/003.

Timeline, Lord Portal archive, Atomic Energy folder B, UK Military Atomic Energy Chronology 1945–1969 (including V bomber force), Portal Collection.

Titterton, Ernest, Secret letter to Allen Brown, secretary, Prime Minister's Department, [early 1953], NAA: A6456, R215/016.

Titterton, Ernest, Secret memorandum to Charles Adams, AWRE, 'Preliminary Results from Group N.M. for T1', undated, Titterton archive, box 100, Series 22, AAS.

Titterton, Ernest, Top-secret report, 'Fast Neutron Measurement at Emu on T-1 and T-2', undated, Titterton archive, box 27, folder 4/89, AAS.

Titterton, Ernest, 'The role of "Observers" and the AWTSC at British Weapons tests, in Australia', 19 May 1985, Titterton archive, box 99, series 22, AAS.

Titterton, Ernest, & Alan Butement, Letter to the editor, sent 19 November 1985, Titterton archive, box 100, series 22, AAS.

Tonkin, David, Transcript of proceedings, Royal Commission into British Nuclear Tests in Australia, NAA: A6448, 15, p. 8632a.

Toon, Erron, Transcript of proceedings, Royal Commission into British Nuclear Tests in Australia, NAA: A6448, 2.

Toon, Erron (Terry), national secretary, Maralinga and Monte Bello Atomic Ex-Servicemen's Association, Letter to Bill Hayden, member for Oxley, 18 May 1984, Malone files.

BIBLIOGRAPHY 331

Top-secret teleprinter message to Flannery, ROCLABS, from Carter, Melbourne, 23 September 1953, DEFE 16/679, NAUK.

Totem Panel, Secret note on Ben Cockram's letter to Allen Brown of 21 January 1953, 6 February 1953, NAA: A6456, R124/018.

Totem Panel, Minutes, Sixth meeting, Swanston Street Melbourne, 26 May 1953, NAA: A6456, R096/002.

Totem Panel, Minutes, Eighth meeting, Swanston Street Melbourne, 24 July 1953, NAA: A6456, R095/012.

Townley, Athol, minister for defence, Cabinet submission no. 466, Nuclear mechanisms – trials at Woomera, 6 November 1959, NAA: A5818, vol. 10/ agendum 466.

Truman, Harry S, Letter to Winston Churchill, 25 March 1951, CHUR 2/28, Churchill Archives.

Tyte, LC, AWRE, Meeting with visiting Australian mission, February 1953, Minutes dated 14 April 1953, AB 16/4544, NAUK.

UK Cabinet Office, Top-secret message to Lord Cherwell's office, 13 October 1953, EG 1/35, NAUK.

UK High Commissioner, Telegram to CRO, 28 February 1953, AB 16/1743, NAUK.

UK Ministry of Defence, Electronic measurements at Tom-Cat, 18 December 1952, DEFE 16/454, NAUK.

UK Ministry of Defence, Secret memorandum, Draft preliminary outline of contribution required by Australian authorities, 12 December 1952, DEFE 16/454, NAUK.

UK Ministry of Defence, Secret memorandum, Totem planning, Outline of site location and layout, DEFE 16/454, NAUK.

UK Ministry of Defence, Secret meeting notes, Totem planning, 19 December 1952, DEFE 16/454, NAUK.

United States Undersecretary for Atomic Energy, Top-secret note from meeting held at St Giles Court London, 28 February 1953, AB 16/1743 TOTEM, NAUK.

Unofficial Background Scientific Information for Press Observers', Briefing notes supplied to Ernest Titterton, [1953], Titterton archive, box 100, series 22, AAS.

Walker, EW, AWRE report no. R67/54, Operation Totem photographic observations, February 1955, ES 5/68. NAUK.

Walsh, RJ, Chairman, Australian Ionising Radiation Advisory Council (AIRAC), letter to JJ Webster, Minister for Science and the Environment, 19 October 1979. NAA: A6456, R057/003.

'Welcome to the Claypan', An introduction to living and working at the Emu claypan campsite, 1953, NAA: D2861, WELCOME TO THE CLAYPAN.

Westwater, FL, commander, Royal Navy, and Freeman, H, Meteorological Office, UK Air Ministry, AWRE Report T12/54 Operation Totem – meteorological services, June 1954, NAA: A6455, RC246.

Williams, Noel, Statements from Australian Witnesses: T–Z, Royal Commission into British Nuclear Tests in Australia, NAA: A6450, 5.

Williams, Noel, Transcript of proceedings, Royal Commission into British Nuclear Tests in Australia, NAA: A6448, 8.

Wilmot, J, minister of supply, Memorandum to Lord Portal, Minutes of the GEN.163 meeting, 10 Downing Street, London, 8 January 1947, Lord Portal archive, Atomic energy folder B, UK Military Atomic energy chronology 1945–1969 (including V bomber force), Portal Collection.

Wilson, Group Captain DA, Report no. T3/54, Canberra Flight Report, Operation Hot Box, ES 5/5, NAUK.
Wilson, JM, Assistant Secretary, Establishments Division, UK Ministry of Supply, Personal and confidential letter to Thomas Elmhirst, 28 January 1953, ELMT 4/2, Elmhirst files, Churchill Archives.
Unofficial Background Scientific Information for Press Observers, Briefing notes supplied to Ernest Titterton, undated but sometime in 1953, Titterton Archive, Box 100, Series 22, AAS.

Secondary sources

Acaster, Ray, 'Sherlock Holmes and the Grim Reaper of Wallatinna: Aboriginal Oral History Solves a Scientific Mystery', *Oral History Association of Australia Journal*, no. 24, 2002.

Arnold, Lorna & Mark Smith, *Britain, Australia and the Bomb: The Nuclear Tests and Their Aftermath*, 2nd edn, Palgrave Macmillan, New York, 2006.

Aylen, Jonathan, 'First Waltz: Development and Deployment of Blue Danube, Britain's Post-War Atomic Bomb', *International Journal for the History of Engineering and Technology*, vol. 85, no. 1, January 2015, pp. 31–59.

Baylis, John, 'The development of Britain's thermonuclear capability 1954–61: Myth or reality?', *Contemporary Record*, vol. 8, no. 1, Summer 1994, pp. 159–74.

Bayly, Ian, *Len Beadell's Legacy: Australia's Atomic Bomb and Rocket Roads*, Bas Publishing, Seaford, 2010.

Berman, R, 'Lindemann in Physics', *Notes and Records of the Royal Society of London*, vol. 41, no. 2, June 1987, pp. 181–89.

Betts, Lewis, *Duncan Sandys and British Nuclear Policy-Making*, Palgrave Macmillan, London, 2016.

Blakeway, Denis & Sue Lloyd-Roberts, *Fields of Thunder: Testing Britain's Bomb*, George Allen & Unwin, London, 1985.

Boyle, Peter G, 'Britain, America and the Transition from Economic to Military Assistance, 1948–51', *Journal of Contemporary History*, July 1987, vol. 22, no. 3, July 1987, pp. 521–38.

Bradshaw, Richard, former Pitjantjatjara Council lawyer, interview with the author, 25 May 2021.

Butler, LJ, 'The Central African Federation and Britain's Post-War Nuclear Programme: Reconsidering the Connections', *Journal of Imperial and Commonwealth History*, vol. 36, no. 3, September 2008, pp. 509–25.

Cane, Scott, 'Something for the Future: Community Development at Oak Valley, S.A.', Report to Oak Valley Community and ATSIC, Ceduna, January 1992.

Cawte, Alice, *Atomic Australia*, New South Wales University Press, Sydney, 1992.

Child, DP & MAC Hotchkis, 'Plutonium and uranium contamination in soils from former nuclear weapons tests sites in Australia', *Nuclear Instruments and Methods in Physics Research B*, 294, 2013, 642–646.

Cowley, Jen with the Uluru Family, *I Am Uluru: A Family's Story*, OMNE Publishing, 2018.

Duguid, Charles, 'The Central Aborigines Reserve', Presidential address, Aborigines Advancement League of South Australia Incorporated, 21 October 1957, Reliance Printing Company, Adelaide, 1957.

Farmelo, Graham, *Churchill's Bomb: A Hidden History of Science, War and Politics*, Faber & Faber, London, 2013.

Fitzgerald, EM, Allison, Attlee and the Bomb: Views on the 1947 British Decision to Build an Atom Bomb', *RUSI Journal*, vol. 122, no. 1, 1977, pp. 49–56.

Fuhrmann, Matthew, 'Spreading Temptation: Proliferation and Peaceful Nuclear Cooperation Agreements', *International Security*, vol. 34, no. 1, Summer 2009.

Gerster, Robin, 'Down the Yellowcake Road: The Minefield of Australian Uranium', *Journal of Australian Studies*, vol. 37, no. 4, 2013, pp. 438–50.

Goldberg, Alfred, 'The Atomic Origins of the British Nuclear Deterrent', *International Affairs* (Royal Institute of International Affairs), July 1964, pp. 409–29.

Goodall, Heather, interview with the author, 10 June 2021.

Goodall, Heather, 'Damage and Dispossession: Indigenous People and Nuclear Weapons on Bikini Atoll and the Pitjantjatjara Lands, 1946 to 1988', prepared for Ann McGrath & Lynette Russell, *Routledge Companion to Global Indigenous History*, 2021.

Heather Goodall, '"The Whole Truth and Nothing But ...": Some Intersections of Western Law, Aboriginal History and Community Memory', *Journal of Australian Studies*, no. 32, March 1992, pp. 104–19.

Goodall, Heather, 'Colonialism and Catastrophe: Contested Remembrance of Measles and Bombs in a Pitjantjatjara Community', in *Memory and History in Twentieth-Century Australia*, eds Paula Hamilton & Kate Darian-Smith, Oxford University Press, Melbourne, 1994, pp. 55–76.

Goodman, Michael S, *Spying on the Nuclear Bear: Anglo-American Intelligence and the Soviet Bomb*, Stanford University Press, Stanford, 2007.

Gott, R, 'The Evolution of the Independent British Deterrent', *International Affairs* (Royal Institute of International Affairs), vol. 39, no. 2, April 1963, pp. 238–52.

Gowing, Margaret, with Arnold, Lorna, *Independence and Deterrence: Britain and Atomic Energy, 1945–1952*, UK Atomic Energy Authority, St Martin's Press, New York, 1974.

Gowing, Margaret, 'James Chadwick and the atomic bomb', *Notes and Records of the Royal Society of London*, vol. 47, no. 1, January 1993.

Hall, S, *Tabloid Man: The Life and Times of Ezra Norton*, HarperCollins, Sydney, 2008.

Hawkins, Dimity, & Matthew Bolton, *Addressing Humanitarian and Environmental Harm from Nuclear Weapons: Monte Bello, Emu Field and Maralinga Test Sites*, International Disarmament Institute and Helene & Grant Wilson Center for Social Entrepreneurship, 2018.

Hill, CN, *An Atomic Empire: A Technical History of the Rise and Fall of the British Atomic Energy Programme*, Imperial College Press, London, 2013.

Hiskey, Garry, *Maralinga: The Struggle for the Return of the Lands*, Wakefield Press, Adelaide, 2021.

Hodges, Barry & Graham Munsell, B Squadron 1st Armoured Regiment (retired), 'The Long Journey', Unpublished profile of the atomic tank, 2019.

Holden, Darren, 'Mark Oliphant and the Invisible College of the Peaceful Atom', PhD thesis, University of Notre Dame Australia, Perth, 2019.

Kerin, Sitarani, '"Doctor Do-Good"?: Charles Duguid and Aboriginal Politics, 1930s–1970s', unpublished PhD thesis, ANU, December 2004.

Krige, John, 'The Peaceful Atom as Political Weapon: Euratom and American Foreign Policy in the late 1950s', *Historical Studies in the Natural Sciences*, vol. 38, no. 1, Winter 2008.

Lapp, Ralph E, 'Sunshine and Darkness', *Bulletin of the Atomic Scientists*, 1959, pp. 27–29.

Lester, Yami, *History and the Land: A Story Told by Yami Lester*, ed. Petronella Water, rev. edn, Institute for Aboriginal Development, Alice Springs, 1990.

Lester, Yami, *Autobiography of Yami Lester*, 2nd edn, Jukurrpa Books, Alice Springs, 2000.

Maclellan, Nic, *Grappling with the Bomb: Britain's Pacific H-Bomb Tests*, Australian National University Press, Canberra, 2017.

McClelland, James, *Stirring the Possum: A Political Autobiography*, Penguin, Ringwood Vic., 1988.

McClellan, Peter, 'Who is telling the truth? Psychology, common sense and the law', *Australian Law Journal*, 2006, vol. 80 (10), p. 655–666.

Milliken, Robert, *No Conceivable Injury: The story of Britain and Australia's Atomic Cover-Up*, Penguin Books, Ringwood Vic., 1986.

Morton, Peter, *Fire Across the Desert: Woomera and the Anglo-Australian Joint Project 1946–1980*, Defence Science and Technology, Department of Defence, Canberra, 1989.

Oak Valley Rangers, 'What, Who, How, Where: A Brief Summary of the OV 2020/21 Oak Valley Ranger Work Program', brochure, Oak Valley, 2020.

Palmer, Kingsley, 'Dealing with the Legacy of the Past: Aborigines and Atomic Testing in South Australia', *Aboriginal History*, vol. 14, no. 1/2, 1990, pp. 197–207.

Pyne, Katherine, 'Art or Article? The Need for and Nature of the British Hydrogen Bomb, 1954–58', *Contemporary British History*, vol. 9, no. 3, 1995, pp. 562–85.

Ruane, Kevin, *Churchill and the Bomb in War and Cold War*, Bloomsbury Academic, London, 2016.

Sherratt, Tim, 'A Political Inconvenience: Australian Scientists at the British Atomic Weapons Tests, 1952–53', *Historical Records of Australian Science*, vol. 6, no. 2, December 1985.

Simpson, John, 'Translating the Art into the Article: Initial British Nuclear Weapon Testing and Production', in *The Independent Nuclear State*: The United States, Britain and the Military Atom, Palgrave Macmillan, London, 1983.

Southwell, Ivan, *Woomera*, Angus & Robertson, Sydney, 1962.

Symonds, JL, *A History of British Atomic Tests in Australia*, AGPS, Canberra, 1985.

Reynolds, Wayne, 'Rethinking the Joint Project: Australia's Bid for Nuclear Weapons, 1945–1960', *Historical Journal*, 41, 3, 1998, pp. 853–73.

Takada, Jun, Hoshi, Masaharu, Sawada, Shozo & Sakanoue, Masanobu, 'Uranium Isotopes in Hiroshima "Black Rain" Soil', *Journal of Radiation Research*, vol. 24, 1983, pp. 229–36.

Theaker, Martin, 'Being Nuclear on a Budget: Churchill, Britain and "Atoms for Peace" 1953–1955', *Diplomacy & Statecraft*, 2016, vol. 27, no. 4, pp. 639–60.

Twigge, Stephen, 'The Atomic Marshall Plan: Atoms for Peace, British Diplomacy and Civil Nuclear Power', *Cold War History*, 16:2, 2016, pp. 213–30.

Tynan, Elizabeth, *Atomic Thunder: The Maralinga Story*, NewSouth, Sydney, 2016.

Vincent, Eve, 'Knowing the Country', *Cultural Studies Review*, vol. 13, no. 2, September 2007, pp. 156–65.

Walker, John R, *British Nuclear Weapons and the Test Ban 1954–1973*, Ashgate Publishing, Farnham, 2010.

Wallace, Phyl & Noel, *Killing Me Softly: The Destruction of a Heritage*, Thomas Nelson, West Melbourne, 1977.

Walsh, Jim, 'Surprise Down Under: The Secret History of Australia's Nuclear Ambitions', *Nonproliferation Review*, Fall 1997.

Walsh, Peter, *Confessions of a Failed Finance Minister*, Random House Australia, Milsons Point, 1995.

Williams, GA, RS O'Brien, M Grzechnik, & KN Wise, 'Estimates of Radiation Doses to the Skin for People Camped at Wallatinna during the UK Totem 1 Atomic Weapons Test', *Radiation Protection Dosimetry*, vol. 174, no. 3, 2017, pp. 322–36.

Williams, GA, Radioactive Waste Safety, 'Maralinga Rehabilitation Project – The Clean-Up at Emu (and What It Means for the Future)', internal, unpublished project summary, Australian Radiation Protection and Nuclear Safety Agency (ARPANSA), 1 August 2017.

Wilson, Denis, & GH Dhenin, 'Some Aspects of Aviation Medicine in Regard to Radiological Hazards', *Proceedings of the Royal Society of Medicine*, 7 October 1954, vol. 48, 1.

Yalata and Oak Valley Communities with Christobel Mattingley, *Maralinga: The Anangu Story*, Allen & Unwin, Sydney, 2009.

Websites

Biographical Dictionary of the Australian Senate, McClelland, James Robert (1915–1999), <biography.senate.gov.au/mcclelland-james-robert/>, accessed 24 September 2021.

Department of Veterans' Affairs, Support for participants in British Nuclear Testing: Compensation and Health Care treatment under the Veterans' Entitlements Act 1986, <www.dva.gov.au/financial-support/compensation-claims/claims-nuclear-test-or-bomb-site-exposure/support-participants>, accessed 8 August 2021.

Eisenhower, Dwight D, 'Address Before the General Assembly of the United Nations on Peaceful Uses of Atomic Energy, New York City', 8 December 1953, <www.presidency.ucsb.edu/documents/address-before-the-general-assembly-the-united-nations-peaceful-uses-atomic-energy-new>, accessed 28 August 2021.

Environmental Protection Agency, Radionuclide Basics: Cobalt-60, <www.epa.gov/radiation/radionuclide-basics-cobalt-60>, accessed 4 July 2021.

8 Expert Committee on the Review of Data on Atmospheric Fallout Arising from British Nuclear Tests in Australia, *Report of the Expert Committee on the Review of Data on Atmospheric Fallout Arising from British Nuclear Tests in Australia* (CB Kerr, chairman; Kerr report), AGPS, Canberra, 31 May 1984, <inis.iaea.org/collection/NCLCollectionStore/_Public/17/019/17019465.pdf>, accessed 26 July 2021.

Mukerjee, Madhusree, 'The Most Powerful Scientist Ever: Winston Churchill's Personal Technocrat', *Scientific American*, August 6, 2010, <www.scientificamerican.com/article/the-most-powerful-scientist-ever/>, accessed 27 August 2021.

Mukerjee, Madhusree, 'Lord Cherwell: Churchill's Confidence Man', HistoryNet, <www.historynet.com/lord-cherwell-churchills-confidence-man.htm>, accessed 5 March 2020.

Obituary for Professor John Carver, 1926–2004, AAS, <www.science.org.au/learning/general-audience/history/interviews-australian-scientists/professor-john-carver-1926-2004>, accessed 9 March 2020.

O'Brien, Anne, 'Hunger and the Humanitarian Frontier', *Aboriginal History*, Volume 39, 2015, <press-files.anu.edu.au/downloads/press/p332783/html/

article05.xhtml?referer=&page=11>, accessed 28 December 2021.

Operation Totem 1953, Nuclear Weapons Archive, <nuclearweaponarchive.org/Uk/UKTesting.html>, accessed 28 January 2021.

'Operation Totem', film produced by the Atomic Weapons Research Establishment, Aldermaston, 1953, <www.iwm.org.uk/collections/item/object/1060036296>, accessed 2 February 2021.

Roitto, Matti, Nevalainen, Pasi, & Kaakoski, Miina, 'Fuel for Commercial Politics: The Nucleus of Early Commercial Proliferation of Atomic Energy in Three Acts', *Business History*, DOI: 10.1080/00076791.2020.1845316, <doi.org/10.1080/00076791.2020.1845316>, accessed 19 July 2021.

The Nobel Prize for Literature 1953, <www.nobelprize.org/prizes/literature/1953/summary/>, accessed 27 August 2021.

Truman, Harry S, 'Statement Announcing the Atomic Bombing of Hiroshima, August 6, 1945', American Experience, <https://www.pbs.org/wgbh/americanexperience/features/truman-hiroshima/>, accessed 27 September 2021.

ACKNOWLEDGMENTS

Researching *The Secret of Emu Field* has been one of the highlights of my life, and I feel grateful and privileged that I have been able to write this story. Of course, no research-based book arrives as the sole creation of the author. Many people have helped me, in a multitude of ways, large and small. Naturally, this book must ultimately be a reflection of my own understanding of the evidence. That understanding has been shaped by knowledgeable people who have generously helped me to find my way through this complex and opaque story. If there are any flaws in my understanding, they are my responsibility alone.

I would firstly like to acknowledge the Aboriginal Traditional Owners of the South Australian atomic test sites. In particular, I would like to thank the Chair of Maralinga Tjarutja Council, Jeremy LeBois, and his fellow Council members, for granting me the right to enter these lands. Also, sincere thanks to Sharon Yendall, who helped facilitate my visit. The Oak Valley Rangers co-ordinators Shane and Samantha Doudle are both amazing, wonderful people. Shane took us to Emu Field, on the bumpy track from Maralinga Village, on a day I will never forget. The caretakers at Maralinga, Roger and Priscilla Petersen, made us feel welcome when we visited, and gave up a huge amount of their precious time showing us around and answering our endless questions. Thank you also to the former Maralinga caretaker, Robin Matthews, who so generously shared his own stories and extensive knowledge.

I hardly know where to begin thanking Andrew Collett, who has been the legal representative for Maralinga Tjarutja for many years, and is knowledgeable and savvy about so many issues connected with the tests. Andrew is also generous, thoughtful, kind and helpful in equal measure. Through Andrew, I met Richard Bradshaw, former Pitjantjatjara Council legal representative, who provided invaluable insights. I was fortunate to be able to speak to the outstanding historian, Heather Goodall, who gave me considerable food for thought; indeed, Heather gave me a new way of looking at some aspects of the story.

Retired ARPANSA scientist Geoff Williams is a fount of knowledge about radiation risks, and has spent a large part of his career looking at what the British left behind them at Maralinga, Emu and Monte Bello. He is also a thoroughly decent person and a great scientific sounding board. I value his scientific knowledge and his generous spirit.

I am nothing if not an archives nerd, and like nothing more than settling into a peaceful archives reading room and opening an old file. This book has taken me to seven archives, all of which have thrilled me. The National Archives of Australia, notwithstanding the well-publicised resourcing problems it has experienced in recent times, is a happy and fruitful place for me. I also love the National Archives of the United Kingdom, housed in its concrete citadel in Kew. I still get a frisson of excitement when I approach that imposing building, walking past ponds graced by peaceful swans. While I am displeased by the secretive impulses of the British government that keep important records about the

British nuclear tests from the eyes of historians and others, I can't fault the UK Archives itself, a highly efficient outfit that (to my continuing amazement) fetches records with dizzying speed. I was fortunate to gain access to the Titterton Archive at the Shine Dome of the Academy of Science in Canberra, a stone's throw from where I used to work at the ANU and a building that first fired my imagination as a kid of seven arriving with my family from Adelaide to live in Canberra. Robyn Diamond at the Academy was so helpful, and my days in the basement working my way through a cache of fascinating Titterton documents live in my memory. I am grateful to Natalie Clough at the Australian Institute of Aboriginal and Torres Strait Islander Studies who helped me to access materials from the Institute's collection. One of the most beautiful libraries I have ever worked in was at Christ Church College at the University of Oxford. The special collections staff there, Cristina Neagu and Alina Nachescu, were just as excited as I was with my findings in the Portal Collection that they helped me access. The Nuffield College Library in Oxford, home to the Lindemann collection, was tranquil and serene, and the archives assistant Emma Quinlan was endlessly helpful. I had a glorious week at the Churchill Archives at the University of Cambridge, where the Director Allen Packwood and his team were welcoming and helpful. Archives are a nation's memory bank. They have not been cherished and cared for enough, certainly in Australia.

Sincere thanks to the investigative journalist Paul Malone, who covered the British tests in Australia extensively during the 1980s. His work for the *Canberra Times*, including with his colleague Howard Conkey, did much to shine an investigative spotlight on the realities of what the British atomic test authorities wrought in Australia. Paul generously gave me access to his extensive collection of original documents on the atomic tests, and I have drawn upon them in parts of this book.

What a fine and intrepid duo of fellow explorers are Lianda Burrows and Darren Holden. These two particularly wonderful people took me seriously when I suggested that we journey together to Maralinga and Emu. We have written about our adventures in a journal article on the landscape of the Maralinga site. Also on that trip were my brother Andrew Burden, who took magnificent photos, and two veterans, Roy Vincent and Barry Hodges. Our little band of travellers had different backgrounds and reasons for visiting the South Australian test sites, but we travelled happily together and enjoyed the ride.

I am so lucky to have met the Maralinga RAF veteran John Folkes and his lovely wife Margaret when I was in the United Kingdom researching this book. They welcomed me into their home in Margate, and have shown me so much love since that visit. John has bravely offered his testimony about the wrongs of the British tests in many different ways.

Thank you to my employer, James Cook University (JCU), for supporting my research. Particular thanks to the Dean of Graduate Research, Professor Christine Bruce, and to Emma Landy, who is a most excellent colleague. Also, I would like to thank the many JCU research degree candidates who, over the years I have worked at the Graduate Research School, have inspired me with their brilliant minds and hearts.

Thank you to the publisher I worked with first at NewSouth, Phillipa McGuinness, who left after contracting this book, and to my new publisher Elspeth Menzies. Both are utterly outstanding publishers, as well as being the best of people. I am indebted to Victoria Chance, a splendid editor who also worked on my previous book, *Atomic Thunder*.

Victoria has an excellent eye and a well-organised mind that she has used to make this book the best it can be. Thanks also to Emma Hutchinson at NewSouth for skilfully steering this book through the production process.

Thanks to Robyn Lewis, Andrew Collett's intern and an ANU law student. Robyn dropped everything when I needed a non-digitised record from the NAA but I was not in a position to hasten to Canberra to get it myself. Thanks also to historian Jacqui Donegan who did the same on another occasion. While I love living in north Queensland, not being able to get quickly to the NAA (especially during a pandemic) does pose particular logistical problems. I would also like to thank the lovely JCU PhD candidate, Ramona Brunner, who kindly provided me with carefully photographed pages of a book she had found that she thought would be of interest to me.

As always, my family have been my most ardent supporters. I will never stop being grateful. Thank you to Mum (Rosemary), Dad (Frank), Inta, Meredith, Andrew, Narelle, Sophie, Alexander and William. Special thanks to Mum, who proofread an early draft with her usual wonderful pedantry, infectious enthusiasm and genuine love of history, and to Andrew, who took brilliant photos of Maralinga and Emu and was such a splendid travelling companion. My wider family also provide me with much love, and I love them back. I mention with sadness my dear Auntie Dod, who died during the final stages of preparation of this book. We also farewelled my beloved Uncle Glen in December 2021. Uncle Glen took special interest in my work on the British nuclear tests. I will miss our conversations.

I have such great friends. I hate to think where I would be without my oldest friend, Susan Davies, who has given me all of the greatest gifts of friendship in full measure. I also mention Nicola Goc, Melissa Lyne, Catherine Naum and Helene Marsh for their kindness, and more. Particular thanks to Julie Funnell, for generous, selfless and extensive assistance during a difficult year, 2021. To my moving picture colleagues and buddies Simon Mathew and Jeff Dawson – one day guys. To JD Mittmann, curator of the art exhibition Black Mist Burnt Country that toured Australia a few years ago, I am grateful for your camaraderie and great discussions over the years about matters atomic. To Brian Toohey, who launched *Atomic Thunder* with a stirring speech and remains one of my hero-journalists, thank you for fighting the good fight. To the many veterans and their families who have contacted me since *Atomic Thunder* came out, I see and hear you, and regret what you have gone through. I see my job as ensuring that the story of what happened can come to light. To Ceri McDade and her colleagues at the British Nuclear Tests Veterans' Association, more power to your arm. Special thanks to BNTVA curator Wesley Perriman for resources such as copies of the magazine *Bulldust* and Emu Field photographs. Both Ceri and Wesley are dedicated advocates for their constituents. My thanks to Stephen Hogeveen, who shared his memories of his Uncle David Bailey, a veteran of the RAAF. Also, Michele Newman helped me to understand the experiences of her father, Morris Newman, also a RAAF veteran.

My community at Alligator Creek has helped me in so many ways, especially in recent times. I mention in particular Jaymie O'Keefe, Alex Rains, Mel and Rob Milla, and Pam Robb, who are the best neighbours imaginable. I love the peace and quiet I am freely able to enjoy these days. Sometimes that peace and quiet is disrupted by my noisy menagerie, but usually only when I am late to feed them. Thanks to Elvis, Adelaide, Fifi, Penguin, Echo, Joanie, Puffin, Palmee, Ash, Jack, Matty,

Junior, Elise and Eddie, for making sure that I keep a reasonable work schedule that does not interfere with feeding duties. To Ramona, Fred, Jane and Lenny, whom I lost recently, you are missed but not forgotten.

Emu Field is in many ways an Aboriginal story, parts of which I am not entitled to tell as a European outsider. I acknowledge the generosity of the Traditional Owners in allowing me to visit and in trusting me to write this book. I do not make any claims to speak for these communities, nor do I presume that my interpretation of events at Emu will match their own. I am humbled by their fortitude in the face of horrendous injustices done to them and I wish them better fortune ahead than they experienced at the time of the tests.

INDEX

Abbreviations:
    Aus.   Australia
    EF   Emu Field test site (X200)
    Maralinga Maralinga test site (X300)
    OH   Operation Hurricane (Monte Bello Islands)
    OT   Operation Totem (Emu Field)
    RAAF   Royal Australian Air Force
    RAF   Royal Air Force
    RC   Royal Commission into British Nuclear Tests in Australia
    TI   Totem I (Emu Field)
    TII   Totem II (Emu Field)
    UK   United Kingdom
    US   United States
    Woomera   Woomera rocket testing range

34 Squadron, RAAF 247
82 Wing, RAAF 168–69, 170, 176, 182

A32 ('High Explosives Research Report No A32') 121, 142, 144–46, 149, 152
A-bombs. *See* atomic weapons
Aboriginal people (Aṉangu) 3
    anxiety 7, 11, 140–41, 157
    civil claim and compensation 126
    deaths 123–25, 157–62
    government and 12, 15–16, 60–61, 123, 258
    health: characteristics 123, 124, 265; problems 131–32, 133–34; psychological 11, 140–41
    land rights 14–15
    lands of 1, 3–4, 5–13, 270
    MacDougall and 10–11, 138–39
    MARTAC clean-up 277
    media and 12, 42, 123
    OT impacts on 4–5, 42, 47–48, 111, 116, 123, 257, 278 (*see also* black mist)
    Penney on 15–16, 261–62
    political consciousness 15
    possible experimentation on 40, 47–48, 94–95
    radiation exposure 124, 127–28, 152
    radioactive contamination 123
    RC 14, 117–20, 125, 130–31, 133–34, 152
    Titterton on 132
    traditional life, decline of 9–10
    water 8, 161
    See also black mist; Oak Valley Rangers
Acaster, Ray 117, 121–22
access (EF) 6–7, 34–35, 49–50, 53–54, 203, 270
accounts of OT, personnel 103–105, 158–61, 162–63

INDEX 345

Adams, Charles 39, 113, 170–71, 244
advance party 33, 49
Advertiser (Adelaide) 17, 94, 123, 132–33, 240, 249–50
aerial surveys 169–70, 171
Age 188, 196–97, 204–205, 206–207
aims (OT) 47, 72, 80, 91, 92–93, 119
air crew. *See* aircraft; aircraft and air crew decontamination
air force operations 165–87
air sampling 165, 166–68, 169, 170, 171
    AWRE 170, 173–74, 176–77
    Penney on 166–67
    TI 170, 173–80
    TII 168, 173–74
Air Staff Requirement notices 68–69
AIRAC (Australian Ionising Radiation Advisory Committee, *formerly* AWTSC) 127, 146–47, 264, 268–69
AIRAC 9 report 127–28, 264, 268
aircraft 50, 54, 58, 59, 86, 151, 164–87
    B29 Superfortress 151, 165, 183–84
    Bristol Freighter 59, 242, 262
    C54 183
    Canberra 164, 165, 173, 174–80
    Dakota 165, 169
    decontamination 168, 170, 173–74, 177, 181, 182, 186
    Hastings 86
    Jindivik 165–66
    Lincoln 164, 165, 168, 169, 170, 171, 172–74, 176, 180–83, 184–86
    Mustang 98
    V bomber 68, 69, 87
aircraft and air crew decontamination 168, 170, 173–74, 177, 181, 182, 186
airliners, commercial 179
airstrip (EF) 5, 54–55, 59, 276
Alexander, AV 65
alpha particles 85

AM (radio program) 132
America. *See* US government
Anangu. *See* Aboriginal people (Anangu)
Anderson, EW 175
Anglo–Australian Joint Project 271, 273, 274
animal life 6, 254
Anne Beadell 'highway' 7
anti-nuclear movement 58, 260–62
Antler, Operation (Maralinga) 4, 63
anxiety 7, 11, 140–41, 157, 163, 184, 279
Argus 196, 225–26
ARL (Australian Radiation Laboratory) 13, 268
Arnold, Lorna 47, 126, 152, 238, 260
'Atom Explosion Success' (*Sydney Morning Herald*) 201–202
'Atomic Bombs, Uranium and Australia's Future' (*Sydney Morning Herald*) 209–210
atomic bunker (UK) 81
atomic clouds 144–45. *See also* Totem I (EF): atomic cloud; Totem II (EF): atomic cloud
atomic energy 87–89, 220, 234
    British: atomic piles 45–47, 87, 232, 234; infrastructure 45–46; Lindemann on 82–83, 88–90; plutonium production 70–73; searches for test sites 43–44
    civilian: Australian uranium and 214, 230, 233, 234; British 46, 66, 74, 83, 87–88, 225
    See also atomic weapons
Atomic Energy Act 1946 (McMahon Act, US) 18, 66, 67, 83, 151, 214
Atomic Energy Agency, International 235z
Atomic Energy Authority (UK) 90, 234
Atomic Energy Commission (US) 219
Atomic Energy Commission, Australian 64, 220

Atomic Energy Research
  Establishment (UK) 66, 174
atomic piles (UK) 45–47, 87, 232,
  234
atomic test sites 19, 43–44. *See also*
  atomic tests in Australia
atomic tests in Australia
  Attlee and 75–76, 77, 78
  Australia: access to information
    153–54, 272–74; attitudes
    50–51; involvement 195
  central desert's suitability for 17
  Christmas Island (Kiritimati) 63
  constraints 65, 71, 73, 80
  costs 254–56
  knowledge gaps 18, 67, 93–94,
    101, 143, 148, 149–50
  Martin and 32–33
  operations: Antler 4, 63, 239;
    Buffalo 4, 87, 157, 239;
    Hot Box 164; Mosaic 4, 63,
    167, 239 (*see also* Operation
    Hurricane (Monte Bello
    Islands); Operation Totem
    (EF))
  risk standards 152
  rockets with nuclear warheads
    271–74
  safety 71–72, 73, 79–80, 92, 144,
    167
  Sandys on 88, 204, 255
  summary 4
  Titterton on 146
  See also Aboriginal people
    (Aṉangu); Royal Commission
    into British Nuclear Tests in
    Australia
atomic war 24, 81–82
atomic weapons
  airborne 79–80, 87
  British: ballistic missiles 87,
    270–71; costs 75, 76, 83;
    Ministry of Supply (UK) and
    48, 76, 88–89; Penney and 58,
    65–66, 73, 79–80, 87, 259;
    Portal and 65; program 18–20,
    32, 51, 58–59, 63, 64, 65–91,
    192–93; secrecy 73, 76, 77,
    153; types 63; unpredictability
    71–72, 73, 79–80, 144 (*see also*
    Blue Danube (Small Boy))
  in conventional warfare arsenal 78
  effects of 148, 199–200
  explosive yields 63, 72
  fuel 46–47
  principles of 63–64, 108–109
  See also atomic tests in Australia
Atomic Weapons Research
  Establishment (UK). *See* AWRE
  (Atomic Weapons Research
  Establishment, UK)
atomic weapons tests in Australia. *See*
  atomic tests in Australia
Atomic Weapons Tests Safety
  Committee (AWTSC, *later*
  AIRAC) 146, 264–66
'Atoms for Peace' agenda (US) 234
Attlee, Clement 65, 66–67, 75–76,
  77, 78
Attorney-General's Department (Aus.)
  242
Australia. *See* Aboriginal people
  (Aṉangu); atomic tests in Australia;
  Australian Government
Australian Army 54, 55–57, 98, 153,
  156
Australian Associated Press 203–204
Australian Atomic Energy
  Commission 64, 220
Australian Defence Committee
  39–40, 215–16
Australian Government
  atomic aspirations 62, 63–64,
    231, 235–36
  Lindemann and 223–28, 232–33
  on media access to tests 205–206
  and minor trials 84
  OH 18, 155, 195, 274
  OT: absence from detonations
    113; access to information 44,
    94, 155–56; approval 25, 44,
    59; British request 24–25,

59, 93–94; costs 254–56;
impatience with British
demands 195–96; involvement
40–41, 195, 208; ministers
notified 29; reimbursement
254, 255, 276; responsibility
for Aboriginal people 15–16,
60–61, 146; secrecy 29, 61,
143, 154, 194–95
security breaches 202–203,
207–208
and test site contamination
267–68
uranium briefing to 213
See also atomic tests in Australia;
nuclear co-operation;
Operation Totem (EF):
Australian involvement
in; uranium: Australian,
negotiations over
Australian Ionising Radiation
Advisory Committee (AIRAC,
formerly AWTSC) 127–28, 146,
264, 268–69
Australian Nuclear Science and
Technology Organisation 277
Australian Radiation Laboratory
(ARL) 13, 268
Australian Security Intelligence
Organisation 202
AWRE (Atomic Weapons Research
Establishment, UK)
Australian mission to 49–50
OT: air sampling 169–170,
173–74, 176–77; black mist
125; fallout 94; film 257;
photographic observation 98;
tanks 98
risk standards 152
AWTSC (Atomic Weapons Tests
Safety Committee, *later* AIRAC)
146, 264–66

B29 Superfortress aircraft 151, 165,
183–84
Bailey, David 158–60

Ball, Rob 132
ballistic missiles 87–88, 271–73
Bayly, Ian 27, 100
Beadell, Len
Bayly on 100
Bulldust on 54
EF: finding and establishment
22, 24, 25–27, 43, 100; maps
4; 'Welcome to the Claypan'
booklet 153–54
Maralinga–Emu Field road 2
OT claims 26–27, 100–101
popularity 26, 100–101
RC 26, 27, 100
Southwell on 26
Stevens, JES, on 99
Vincent on 25
Woomera 26
Beale, Howard
on Australia and atomic energy
220
on Australian security
arrangements 202, 203
correspondence: Churchill
219–21; Menzies 239; Spender
212
on EF permanency 239
OT 32, 203, 208
on servicemen and exaggeration
185
on souveniring 244
on uranium negotiations 220–21
beryllium 84–86
Bevin, Ernest 65
Bikini Atoll tests 190, 204
Black, DH 249
black mist 92, 117, 120–41
blast pressure 107
Blue Danube (Small Boy)
fuel 72
Kittens trials 85, 86
models 18, 93
OH 18, 58, 93, 144
OT 80, 93, 144
production ended 58
RAF and 72, 81

service ended 272
uses 87, 197, 272
Blunden, WR 105–106
bomb towers 13, 57, 93, 129
Bowen 111
Bradshaw, Richard 136–37
Brinkley, Tony 103–105, 108
Bristol Freighter aircraft 59, 244, 264
Britain. *See* atomic energy: British; atomic tests in Australia; atomic weapons: British; British government
British Army 97, 101–102
British government
    atomic war, preparing for 81–82
    and media access to tests 188–89, 205–206
    OT 23–24, 61, 93–94, 113, 254–56, 276
    scrutiny, lack of 50, 51
    secrecy 19–21, 73, 76, 77, 152–53, 257–58
    on security and media breaches 209–10
    US government and 17–18, 74–76, 78, 90–91
    See also atomic tests in Australia; nuclear co-operation; uranium: Australian, negotiations over
British scientists 99–100
Brooking, P 48
Brown, Allen
    correspondence: Cockram 28, 113, 114; Menzies 113, 114, 208, 227, 230; JES Stevens 113; Titterton 30
    OT 113–14, 208
    on uranium negotiations 226, 230
    on Woomera Prohibited Area extension 28
Buffalo, Operation (Maralinga) 4, 87, 129, 157
Bulldust (EF newsletter) 54
Bulletin 133, 226
bunker, atomic (UK) 81
Butement, Alan

Australian Defence Committee meeting 40
EF 24, 27, 43, 225
OT 99, 195, 208
Penney and Martin meeting 22–23
Titterton and 266

C54 aircraft 183
Cabinet Committee on Uranium (Aus.) 226–27
Calder Hall atomic piles (PIPPAs) 46–47, 87, 232, 234
camera towers 99
Cameron, James 188, 204–205
Canada 44
Canberra, UK High Commission in 194
Canberra aircraft 164, 165, 173, 174–80
Canberra Times 42, 173
cancers 173, 181, 275
Caplehorn, WF 105–106
Carter, Leo 35, 49
Carver, John 108, 109
Casey, Richard 213
Cawte, Alice 225
CDA (Combined Development Agency) 214, 216–17, 225, 233
central desert 8, 17
Centurion tank 36, 54, 98, 240
Chadwick, James 69–73, 74, 215
Cherwell, 1st Viscount. *See* Lindemann, Frederick (1st Viscount Cherwell)
Chiefs of Staff (Aus.) 113
Chiefs of Staff (UK) 75, 81, 97, 113
Child, DP 269–70
Chilton, Fred 40
Christmas Island (Kiritimati) tests 63
Churchill, Winston
    and Britain's atomic weapons program 77–78, 79
    correspondence: Beale 219–21; Lindemann 77–78, 82–84, 89–90, 188–89, 216, 217–19,

223–25; Menzies 112, 228–29, 233
Eisenhower and 82–83
Lindemann and 67–68, 77
on OT fallout 115–16
on Penney 259
and Quebec Agreement 74
Ruane on 79
and thermonuclear weapons 79–81
on TI 112
civil claims 126
civil service (UK) 88–90
Clark, Allen 184
Clarke, John 186
claypans
  Dingo 27, 166
  EF 5, 53, 59, 175, 276
clean-up operations 4, 275, 276–77
co-operation, nuclear. *See* nuclear co-operation
cobalt-59 129
cobalt-60 129–30, 268
Cochrane, Ralph 79
Cockcroft, John 59, 66, 69, 151
Cockram, Ben
  correspondence: Brown 28, 113, 114; McKnight 190, 191; Menzies 28–29, 143, 189–90; Pritchard 28, 60–61, 206, 231–32
  on EF meteorological conditions 28
  on media 189–90, 206
  OT 29, 113, 114, 143
Collett, Andrew 125, 145–46
Collins, John 39–40
colonialism 118, 263
Colquhoun, David 169, 171, 172
Combined Development Agency (CDA) 214, 216–17, 225, 233
commercial airliners 179
commercialisation of nuclear energy 88–89
Commonwealth Investigation Service 37, 245

Commonwealth Meteorological Service 111
Commonwealth Relations Office 44, 156–57
Commonwealth resources 83, 216–17, 218
community memories 137–40
compensation payments 126, 173, 185, 186
conditions (EF) 33–34, 37–39, 45, 49, 55–56, 57, 100, 241–43
Conkey, Howard 42
construction (EF) 5, 33–34, 54–56
contaminated waste 50
contamination, radioactive. *See* radioactive contamination
Coober Pedy 140
Cook, William 101
cooperation, nuclear. *See* nuclear co-operation
Coronation, Elizabeth II 51, 52, 56–57
Cosgrove, Roy 164, 184–85
costs, atomic tests 75, 76, 83, 240, 254–56, 276
court martial 244–49
cover stories 28, 53, 153, 196
Crimes Act (Aus.) 194, 275
Crofts, JM 111
Cullinan, Tom 118–19
Cutter, Trevor 123, 133

D-notice system 190–91, 196, 205
Daily Mail (London) 207
Daily Mirror (Sydney) 198, 227
Daily Telegraph (Sydney) 142, 210
Dakota aircraft 165, 169
Daley, EA (Ted) 182–83
Davey, George 194, 195–96
Dead Heart, Operation (EF) 170
deaths 121, 123–25, 127, 157, 158–60, 264
decontamination, aircraft and air crew 168, 170, 173–74, 177, 181, 182, 186
Defence, Department of (Aus.) 6–7, 270

Defence, Ministry of (UK) 48, 136
Defence (Special Undertakings) Act 1952 (Aus.) 34–35
Defence Committee (Aus.) 39–40, 215–16
Defence Committee (UK) 82
Defence, Press and Broadcasting Committee (Aus.) 205
Defence Research and Development Policy Committee (Aus.) 62, 64
Defence Sub-Committee (UK) 75
Defence White Paper (Sandys) 79
Department of Defence (Aus.) 6–7, 270
Department of Supply (Aus.) 29–30, 53, 242, 245
detonations. *See* Totem I (EF): detonation; Totem II (EF): detonation
Dhenin, Geoffrey 164, 175–77, 178–80
Dingo claypan 27, 166
Dolarinna (Larry's Well) 7
dosimeters 177
Drake Seager, E 20–21, 106–107
Duguid, Charles 9–10
Dvoretsky, Harold 192–93

Eames, Geoff 125–26, 141, 145–46
Earner, Mark 162–63
Eastern Area Detachment Woomera 182
Edelman, Maurice 232
Eden, Anthony 79
Edwards, Lancelot 172–73
Egan, LG 34
Eisenhower, Dwight 82–83, 230, 234
electromagnetic pulse measurements 151
Elizabeth II 51, 52, 57
Elmhirst, Thomas 22, 51–52, 53, 97, 100, 108, 150, 207–208
emphysema 253
Emu Field–Maralinga road 2
Emu Field test site (X200)
  access 6–7, 35–36, 49–50, 53–54, 201, 272
  airstrip 5, 37, 54, 59, 176
  conditions 33, 37–39, 45, 49, 55–56, 57, 101, 244–45
  construction 5, 33–34, 54–56
  costs 240, 254–56
  descriptions 1–7, 12–13, 27, 50, 245–46, 250
  equipment 36–37, 49, 101, 240–42, 244–50, 251, 254, 255
  evacuation 239–43, 248
  finding and establishment 19, 22–24, 25–28, 43, 58–59, 100
  government files on 19–21, 257–58
  legacy 257, 279
  Penney and 22, 24, 27, 44–45, 58–59, 85
  permanency 28, 47, 239
  Redmond on 39
  Register of the National Estate 276
  RC on 42, 239
  secrecy about 28–29, 37, 43, 143, 191–92, 194–96, 243, 257–58
  Southwell on 12–13
  statements to media about 52–53
  suitability 5, 19, 22, 27–28, 42–43, 238–39
  Supply, Department of (Aus.), and 29–30, 242–45
  Supply, Ministry of (UK), and 21
  Woomera, conflated with 196–97
  See also Kittens trials; Maralinga Tjarutja lands; Operation Totem (EF); Totem I (EF); Totem II (EF)
Emu report (AIRAC) 268–69
equipment (EF) 36–37, 49, 101, 240–42, 244–50, 251, 254, 255
Ernabella Mission 9–10, 11, 137–40
establishment (EF). *See* Emu Field test site (X200): finding and establishment

europium-152 268
evacuation (EF) 239–43, 248
executive committee (OT) 48
experiments
　categories of 30–31
　OH 99, 106, 107, 149, 167, 172, 182
　OT: military 44–45, 97–99, 101–102, 105–106, 164–65, 169–71; scientific 96, 107–110
　See also RAAF (Royal Australian Air Force): OT; RAF (Royal Air Force): OT; *individual types of experiments*
Expert Committee on the Review of Data on Atmosphere Fallout Arising from British Nuclear Tests in Australia, report (Kerr report) 124, 168, 263
explosions. *See* Totem I (EF): detonation; Totem II (EF): detonation
explosive yields
　atomic weapons 63, 72
　inaccuracy of British predictions about 58, 108
　measurements 167
　OT 144
　thermonuclear weapons 63
　TI 57, 94, 103, 107, 145, 152
　TII 58, 94, 107

Fadden, Arthur 25, 29, 205–206
fallout
　OH 94, 95, 102, 264, 272
　OT 94–96, 102, 115–16, 126, 152, 156, 238, 264, 274
　RC on 152
　TI 94–95, 115, 126
　TII 94–95, 115, 238
Farmelo, Graham 259–60
fast neutrons 31, 108–110, 129
film badges (EF) 98, 105–106, 177–78, 251
finding and establishment (EF) 19, 22–24, 25–28, 43, 58–59, 100

fission 46–47, 63
fission weapons. *See* atomic weapons
Fitch, Jill 14
Fort Halstead 49
Fourth Estate 189
Francis, Desmond 244–46, 247
Fraser, Malcolm 15, 267
Fuchs, Klaus 22
fuel, weapon 119
fusion 63
fusion weapons. *See* thermonuclear weapons

Gallie, G 86
Garden (squadron leader, Aus.) 262
Gates, BG 49
GEN.163 committee 65–66, 73
Gerster, Robin 236
Geschke, Charles 181–82
Gibbs, William 265–66
Giles, Ellen 135
glass. *See* trinitite
Gold Card medical assistance 187
Goodall, Heather 10, 116, 121, 137–40, 161
Goodman, Michael 151
government. *See* Australian Government; British government; US government
Gowing, Margaret 78, 89, 215
Granite Downs settlements 12
Great Victoria Desert 8
ground zeroes 13, 102, 268
Groves, Leslie 72

H-bombs. *See* thermonuclear weapons
Hamilton, Annette 120
Hardman, Donald 40
Hastings aircraft 86
Hattam, E 250, 251
Hawke, Bob 267
Hayden, Bill 158
Hayes, EA (range security officer) 253–54
health characteristics, Aboriginal people 123, 124, 265

health problems
  in Aboriginal people 11, 131–32, 133–34
  compensation for 173, 185, 186
  debates over 264–65
  government assistance for 181, 187
  in military personnel 169, 173, 181, 274
  in peace and security officers 252–54
heat of detonations 12, 106, 107
helicopters 97
Herald (Melbourne) 193–94, 197, 223
HEROD Committee 79, 80–81
High Commission in Canberra (UK) 194
High Commission in New Zealand (UK) 156–57
High Explosive Research. *See* High Explosives Research (HER, UK)
High Explosives Research (HER, UK) 45, 89, 150. *See also* 'High Explosives Research Report No A32'
'High Explosives Research Report No A32' 121, 142, 144–46, 149, 152
Hill, CN 46
Hill, JF, Report of enquiry into souveniring 237, 245–50
Hinton, Christopher 69
Hogeveen, Stephen 158–60
Hot Box, Operation (EF) 164
Hotchkis, MAC 269–70
How, FC 48, 97, 155
Hungerford, 1st Viscount Portal of (Charles Portal) 65–70
HUREX 48
Hurricane, Operation (Monte Bello Islands). *See* Operation Hurricane (Monte Bello Islands)
hydrogen weapons. *See* thermonuclear weapons

inaccurate predictions 94, 144–45, 149, 152
India 140
information, public 29, 153, 154, 156, 208
initiators 84–85
International Atomic Energy Agency 235
International Commission on Radiological Protection 152

Jindivik aircraft 165–66
Joint Chiefs of Staff (UK) 97
Joint Defence Facility, Pine Gap 235
Jonas, William 14

Kanytji 134
Kerr, Charles 124. *See also* Kerr report
Kerr report 124–25, 168, 263
keys, personnel 154
Kiritimati (Christmas Island) tests 63
Kittens trials 4, 49, 84–86, 203–204
knowledge gaps, British 18, 67, 94, 102, 144, 149, 150–51
Kunmanara 137–39

land rights movement 14–15, 136
Lander, Merdie 135–36
Larry's Well (Dolarinna) 7
Lashmar, Paul 133
legacy (EF) 257, 279
Leigh, David 133
Lennon, Lallie 134–35
Lester, Karina 257
Lester, Lucy 136
Lester, Yami 1, 7–8, 11, 14, 117, 131–34, 136–37, 267
lethal ranges, atomic weapons 109, 160
leukaemia 169
lies about OT 143
Lincoln aircraft 164, 165, 168, 169, 170, 171, 172–74, 176, 180–83, 184–86
Lindemann, Frederick (1st Viscount Cherwell) 67–68

Australia visit 221, 224–27, 232–33
Australian Government, negotiations with 223–28, 232–33
  on Britain's atomic future 82–83, 88–90
  correspondence: Churchill 78, 82–83, 89–90, 188–89, 216, 217–19, 223–25; Elmhirst 51; Menzies 24–25, 218, 254; Ministry of Supply (UK) 94; Penney 24
  Churchill and 67–68, 77
  and Defence Committee 82
  Menzies on 233
  OH 78
  OT: absence from 113; on Australian access to information 94; importance to 89; on media access 113, 188–89; naming 12; on reimbursement of Australian Government 254; request to Australia 24–25
  paymaster-general 68
  on Penney 259
  Sandys and 221
  and uranium 83, 209, 216–17, 218
  on US nuclear co-operation 90, 216, 217–18
living creatures 6, 254
Lloyd, Bill 26, 27
Lloyd, FB 44, 249
Long Range Weapons Establishment (LRWE, Aus., *later* Weapons Research Establishment) 23, 243, 251
Lowery (major, Aus.) 248, 262
LRWE (Long Range Weapons Establishment, Aus., *later* Weapons Research Establishment) 23, 243, 251
Lucas, Leonard 49, 50, 55–56, 101, 246, 262

Mabel Creek 12
Macaulay, Robert 11
MacDougall, Walter 5–6, 10–11, 61, 118–19, 130, 138–39
Macmillan, Harold 79
Mail (Adelaide) 271
mail (EF) 243, 246–48
Makins, Roger 59
Malone, Paul 42
Man to Man (Menzies, radio program) 200
Manhattan Project 18, 62, 67, 76
Maralinga–Emu Field road 2
Maralinga Rehabilitation Technical Advisory Committee (MARTAC) 96, 277
Maralinga test site (X300) 3, 4, 63, 84, 86, 87, 116, 129, 268, 269, 275
Maralinga Tjarutja 3, 7, 15, 270, 277–78. *See also* Maralinga Tjarutja lands
Maralinga Tjarutja Land Rights Act 1984 (Aus.) 3
Maralinga Tjarutja lands 1, 3–4, 5–12, 270
Marla Bore 14, 118–20
MARTAC (Maralinga Rehabilitation Technical Advisory Committee) 96, 277
MARTAC report 96
Martin, Leslie
  and atomic tests in Australia 32
  on atomic weapons' effects 147
  Australian Defence Committee, meeting with 39–40
  correspondence: Menzies 148–49; Penney 149–50
  McKnight on 32, 149, 194
  OH 194–95
  OT: access to information 154; advice 39–40; involvement 99, 194–95, 208; reliability of predictions 148; safety evaluation 147–50
  Penney and Butement, meeting with 23

Penney on 32
Mary Kathleen uranium deposit 214
materials (EF). *See* equipment (EF)
McBride, Philip 25
McClellan, Peter 96, 161–62, 163, 166–67, 168–69, 267
McClelland, James (Diamond Jim) 14, 19, 30, 118–19
McEwen, John 272–73
McKnight, Allan
   correspondence: Cockram 190, 191; Davey 194, 195–96; Shedden 195; JES Stevens 207–208
   on Martin 32–33, 149, 194
   on media access 190, 191
   request for secrecy 195–96
   on Titterton 32–33, 149
McMahon Act (Atomic Energy Act 1946, US) 18, 66, 67, 83, 151, 214
measles epidemic, Ernabella Mission 11, 137
media
   access to atomic tests 113, 188–89, 190, 191, 205–206
   D-notice system 190–91, 196, 205
   statements received: Aboriginal inhabitants 12; minor trials 204; OT 52–53, 55, 156, 205; RAAF's involvement 171; Titterton's involvement 195; TI 200–201, 209; from Totem Panel 53, 55, 191–92
   OH and 188, 190–91
   OT and 113, 188–211
   reports published: Aboriginal people 42, 123; ballistic missiles 271; black mist 132–33; Edwards 173; EF evacuation 240; Menzies 198–99; minor trials 203–204; misinformation in 192–93, 194, 196–98, 199; OT 192–93, 203, 207, 208; public safety 198–200, 210; radiation exposure 199–200; Sandys' opinion of Woomera 223; security 142, 202–203, 210; South Australia's suitability for atomic tests 17; souveniring 249–50; suppression of 52; TI 92, 188, 201–202, 204–205; TII 203; uranium negotiations 209–210, 222–23, 225–26, 227, 235
   at TI 189, 200–201, 208–209
medical expenses, assistance for 181, 187
Melbourne 148
memories, community 137–40
Menzies, Robert
   on Australian security breaches 203
   correspondence: Beale 239; Brown 113, 114, 208, 227, 230; Churchill 112, 228–29, 233; Cockcroft 59; Lindemann 24–25, 218, 254; Martin 147, 148–49; Salisbury 229
   on Lindemann 233
   Man to Man (radio program) 200
   media reports on 198–99
   OH 18
   OT 147, 148–49, 198, 200, 206–207
   Sandys and 222
   on TII 203
   UK visit 223–24
   uranium negotiations 218, 219, 222, 223–24, 225–26, 227, 228–29
   US visit 219
Messel, Harry 199
meteorology
   EF 28, 30–31, 45, 111–12
   OT 96, 111–12, 145–46, 149–50, 265
   TI 12, 92, 96, 121–22, 145–46
   TII 96, 114–15, 238
military experiments 44–45, 97–99,

101–102, 105–106, 164–65, 169–71. *See also* RAAF (Royal Australian Air Force): OT; RAF (Royal Air Force): OT
military observers 113–14
military operation 39–40, 49, 186
military planning (UK) 81–82
Ministry of Defence (UK) 48, 137
Ministry of Supply (UK). *See* Supply, Ministry of (UK)
minor trials 4, 49, 84–86, 203–204
Mintabie 12, 119, 121, 134
miscarriages 253
misinformation in media reports 192–93, 194, 196–98, 199
missiles, ballistic 87–88, 271–73
Moffitt, Frank 250–51, 252, 254
Monaghan, JG 98
Monson, Ronald 142, 201, 210
Monte Bello Islands test site 22, 43, 44, 45. *See also* Operation Hurricane (Monte Bello Islands); Operation Mosaic (Monte Bello Islands)
Morgan, Frederick 97
Morning Bulletin (Rockhampton) 235
Moroney, John 263–64
Morrison, CS 251–52
Morrison, Herbert 65
Mosaic, Operation (Monte Bello Islands) 4, 63, 167
Murray, Thomas 253, 254
Mustang aircraft 98

Naggs, Rex 168–69
National Archives (UK) 20–21
native patrol officers 10–11, 61
Nettley, Richard 183
neutrons 31, 46, 84–86, 108–110, 129, 268
Nevada nuclear tests 189
Never Never 135
New Zealand 156–57
Newall (sergeant, peace officer) 37, 250, 251
Newman, Michele 181
Newman, Morris 180–81

Non-Proliferation Treaty (Treaty on the Non-Proliferation of Nuclear Weapons) 235
nuclear co-operation 18–19, 22–23, 83, 90, 207, 212–36, 272. *See also* McMahon Act (Atomic Energy Act 1946, US); Quebec Agreement
nuclear energy, commercialisation of 88–89
nuclear weapons. *See* atomic weapons; thermonuclear weapons
numbers of personnel (EF) 22, 59

Oak Valley 6
Oak Valley Rangers 5, 6, 278
Observatory Hill 1, 2
Observer 133
observers 99, 113–14, 147, 195, 201–202, 208, 209
O'Connor, Frank 255
Official Secrets Act (UK) 52, 160, 194, 275
O'Grady, Francis 49
Oliphant, Mark 62, 213, 232, 260
Olney (lieutenant, Aus.) 244
Operation Antler (Maralinga) 4, 63
Operation Buffalo (Maralinga) 4, 87, 129
Operation Dead Heart (EF) 170
Operation Hot Box (EF) 164
Operation Hurricane (Monte Bello Islands)
    Australian Government and 155, 195, 274
    costs 40–41, 76
    Defence (Special Undertakings) Act 1952 and 34–35
    experiments 101, 106, 107, 149, 167, 172
    fallout 94, 95, 102, 264, 274
    HUREX 48
    Martin and 194, 195
    media and 188, 190–91
    Menzies' approval of 18
    Penney on 22
    RAAF 182

RAF 167
risks 78, 151, 172
Royal New Zealand Air Force 156–57
safety 146, 167, 182, 186, 263–64
site 4, 22, 44–45, 58, 96–97
Solandt on 24
success of 17–18, 22, 24, 80, 82, 90–91
technology 63
Titterton and 31
weapon tested 18, 58, 93, 144
Operation Mosaic (Monte Bello Islands) 4, 63, 167
Operation Tom-Cat. *See* Operation Totem (EF)
Operation Totem (EF)
accounts of, personnel 103–105
aims 47, 72–73, 80, 91, 92–93, 119
Australian involvement in: access to information 29, 93–94, 143, 151, 153, 154–55, 194; AIRAC on 146–47; Australian Army 153, 156; Australian Government 40–41, 195, 208; Butement 99, 195, 208; Martin 99, 194–95, 208; Menzies on 147, 200; observers 99, 113–14, 147, 195, 200–201, 208; Penney on 155, 261, 262–63; Titterton 30–33, 99, 108–110, 146, 195, 208 (*see also* RAAF (Royal Australian Air Force): OT)
Beale on 32–33, 202–203, 208
costs 240, 254–56
executive committee 48
Goodall on 116
name 12
summary 3–4
transportation of bombs 86
See also Emu Field test site (X200); Totem I (EF); Totem II (EF)
Owen, Leonard 67

Pacific Islanders 63, 258
Palmer, Kingsley 118
Parker (flight lieutenant) 86
patriotism 50–51, 56–57, 191, 193–94
Pattie, Geoffrey 136
Peace Officer Guard 241
peace officers 37–39, 250–54
Pearce, Noah 99, 268, 275
Pearce report 268, 275
Penney, William
on Aboriginal people 15–16, 262
on air sampling 166–67
anti-nuclear turn 58, 258–60
and atomic weapons 58, 65–66, 73, 79, 87, 258
on Australian personnel 261–63
on Beadell 26
on black mist 126
Blue Danube 58
correspondence: Cochrane 79; destruction of 260; Elmhirst 97; How 48; Lindemann 24; Sandys 222; JES Stevens 26, 111, 260–62; Titterton 149–50
EF 22, 24, 27, 44–45, 58–59, 85
Lindemann on 259
on Martin 32
meetings: Butement and Martin 22–23; Ministry of Supply (UK) 52
OH 22
OT 260–64; Australian involvement 155, 261, 262–63; constraints 97; executive committee 48; fallout 94, 95; military demands 99, 101; observers 99, 209; safety 143–46, 148, 49, 261–62; secrecy 144; security 35–36; urgency 94
Portal committee 69
RC 15–16, 40, 44–45, 85, 95, 96, 126, 145–46, 166–67, 260
Titterton and 31, 32, 146, 263

TI 102, 103–104, 111, 146, 209
Tube Alloys 58
permanency (EF) 28, 47, 239
permits, visitor 6–7, 270
Perrin, Michael 69
personnel accounts (OT) 103–105
personnel keys 154
personnel numbers (EF) 22, 59
photography 99, 170–71, 209
Pingkai (Pingkayi) 134
PIPPAs (pressurised piles producing industrial power and plutonium, Calder Hall) 46–47, 87–88, 232, 234
Pither, Alfred 49
Playford, Tom 220
plutonium 70–73, 83–84, 88, 212, 268, 269
plutonium-239 45–46, 70, 72, 270
plutonium-240 46, 47, 71–72, 269, 270
polonium-210 84, 85, 86
Portal, Charles (1st Viscount Portal of Hungerford) 66–70
predictions, inaccurate 94, 144–45, 147, 151–52
press. *See* media
pressurised piles producing industrial power and plutonium (PIPPAs, Calder Hall) 46–47, 87–88, 232, 234
Prime Minister's Department (Aus.) 190
Pritchard, Neil 21, 60–61, 206, 231–32
Proceedings of the Royal Society of Medicine 179
production methods, plutonium 70–73
progress reports 56–57
propaganda 189
public information 29, 153, 154–55, 155, 208
Pugwash 260
Purdie, AC 103

Quebec Agreement 74–76, 82, 214

RAAF (Royal Australian Air Force)
34 Squadron 247
82 Wing 168–69, 170, 176, 182
accounts of members 157–60, 161–62
aerial surveys 169–170, 171
compensation 173, 185, 186
Coronation celebration 56–57
EF 5, 33, 49, 54, 241–42
medical expenses, assistance for 181, 187
OH 182
OT 165, 168–74, 176, 180–83, 184–87
Penney on 262
radiation exposure 168
radioactive contamination 168–70, 171–72, 181–83, 184–87
risks to personnel 165, 168, 172, 182, 185–86
special leave for personnel 56
Totem Panel 169
radiation exposure
Aboriginal people 123–24, 127–28, 129, 152
Australian Army 98
EF 254
media reports on 199–200
OT 164–65, 182–83, 264
RAAF 168
RAF 176
TI 98, 104–106, 164–65, 167–68, 170, 172, 176–78, 180, 183–84
TII 173
See also radioactive contamination
radiation monitoring
OH 167
OT 109, 111, 147, 183–84
TI 104–106, 109, 167–68, 183–84, 185–86
radioactive contamination
Aboriginal people 127

EF 13, 85, 243, 251, 252–53, 268–69, 275
Maralinga 268, 269
minor trials 85
OT 28, 50, 95–96, 142, 145, 150, 164–65, 168–71, 268–70
RAAF 168–70, 171–72, 181–83, 184–87
RAF 164–65, 175–80
TI 95–96, 105, 111, 122, 126–30, 167–70, 171–78, 180, 183–84, 268–70
TII 95, 111, 114–15, 168, 173–74, 238, 268–70
See also radiation exposure
'Radiological Safety and Future Land Use at the Maralinga Atomic Weapons Test Range' (AIRAC) 268
radiological safety orders 172
radium 214
Radium Hill uranium deposit 214, 222, 223
RAF (Royal Air Force)
  airborne atomic weapons 79–80
  OH 167
  OT 165, 174–80
  radiation exposure 176
  radioactive contamination 164–65, 175–80
  risks to personnel 179–80
  See also air force operations
range security officers 37
reconnaissance (EF). *See* finding and establishment (EF)
Redmond, TWJ 39
Register of the National Estate 276
rehearsals 87
Report of enquiry into souveniring (Hill) 237, 244–46, 247, 248–49
reports, media. *See* media: reports published
resources, Commonwealth 83, 216–17, 218
Resources and Energy, Department of (Aus.) 183
risk standards 151–52

risks
  OH 78, 151, 172, 182
  OT 144, 150, 152, 165, 168, 172, 179–80, 182–83, 185–86
  See also safety
Rivett, David 213
Riviere, AC 108
rockets (OT) 102–103, 178
Roosevelt, Franklin 74
Rose (wing commander, Aus.) 164
Rowell, Sydney 39
Royal Air Force. *See* RAF (Royal Air Force)
Royal Australian Air Force. *See* RAAF (Royal Australian Air Force)
Royal Commission into British Nuclear Tests in Australia
  Aboriginal oral evidence 14, 117–20, 131–32, 133–34
  British evidence 15–16
  establishment 13–14
  evidence: Beadell 26, 27, 100; Clark 184; Collett 145–46; Colquhoun 171; Cosgrove 184–85; Eames 125, 141, 145–46; Earner 162–63; Edwards 172, 173; Geschke 181–82; Giles 135; Hamilton 120; Kanytji 134; Lander 135–36; L Lennon 134–35; Y Lester 117, 134; Moroney 263–64; Murray 254; Naggs 168–69; Morris Newman 181; Penney 15–16, 40, 44–45, 85, 95, 96, 126, 145–46, 166–67, 260; Pingkai (Pingkayi) 134; Siddons 95; Titterton 147; Tonkin 131–32; Toon 157–58; N Williams 252–53; Wise 264
  findings and comments: A32 152; Aboriginal deaths 125; Aboriginal radiation exposure 152; AIRAC 9 report 127, 264; AWTSC 266; black mist 126, 131; EF 42, 239; fallout 152; media access 189; minor

trials 84; radioactive exposure 168-69, 170, 171-72, 185-86, 274-75; risk standards 151-52; safety 150-51, 165; secrecy 19; TI 152; TII 114-15; UK 120 history of tests commissioned for 183-84
McClellan: and oral evidence 161-62, 163; questions 96, 166-67, 168-69; Titterton on 267
Titterton on 264-67
Royal New Zealand Air Force 156-57
Ruane, Kevin 79, 83
Rum Jungle uranium deposit 214, 216, 219, 235
rumours 157, 240
Russia (Soviet Union) 62, 77, 81, 166-67

safety
British recklessness 92
media reports on 198-200, 210
OH 146, 167, 182, 186, 263-64
OT 143-54, 165, 206, 261-62
radiological safety orders 172
RC on 149-50, 165
Totem Panel on 148
Salisbury, marquis of 229
Sandys, Duncan
on atomic tests 88, 204, 255
Australia visit 221-23, 230-31
Defence White Paper 79
Lindemann and 221
meetings: Penney and Elmhirst 53; Playford 222-23
Menzies and 222
OT 111-12, 114, 222
Penney, correspondence with 222
request to test rockets with nuclear warheads 271-74
on uranium 215
on Woomera 223
schedule (OT) 84
scientists, British 99-100
Scotland 81

scrutiny, British government 50, 51
secrecy
Australian Government 29, 61, 143, 153, 154-55, 194-95
British government 19-21, 73, 76, 77, 152
cover stories 28-29, 52-53, 143, 153, 194
EF 28-29, 37, 43, 141, 191-92, 195-96, 243, 251-52, 257-58
McKnight 195-96
media 155 (*see also* D-notice system; media: access to atomic tests)
Menzies 198, 206-207
Penney 144
RC on 19
TI 202, 210
TII 92, 112-13
See also security
security
Beale on 202, 203
breaches of 202-203, 207-208
EF 37, 39, 141, 250-54
media reports on 143, 202-203, 210
OT 21, 22, 35-36
See also secrecy
Shedden, Frederick 59, 195
shock waves 96
Siddons, Edward 95
Sixty Minutes (television program) 185
slit trench measurements 106-107
slow neutrons 108, 109, 110
Small Boy. *See* Blue Danube (Small Boy)
Smith, Mark 47, 126, 152, 238
Solandt, Omond 24, 39-40
South Australia 17, 220, 222-23
South Australian Health Commission 123-24, 268
Southwell, Ivan 12-13, 26, 238
souveniring 237, 244-50
souveniring, report of enquiry into (Hill) 237, 244-46, 247, 248-49

Soviet Union 62, 77, 81, 166–67
special leave, military personnel 55–56
Spender, Percy 212
Stafford-Cripps, Richard 75
statements, media. *See* media: statements received
Stevens, Donald 266
Stevens, Duncan 62
Stevens, JES (Jack) 61–62
    on Beadell's site establishment 100
    on British authorities and media leaks 207–208
    correspondence: Brown 113; Elmhirst 100, 108–109, 149–150, 207–208; McKnight 207–208; Penney 26, 111, 260–62
    and Melbourne concerns 149
    and Titterton's experiments 108
Stinson, CA 237
stories. *See* cover stories; rumours
Strauss, George 75–77
Stubbs Walker, J 207, 208
suitability (EF) 5, 19, 22, 29–30, 42–43, 238–39
Sun (Sydney) 198
Sunday Herald (Sydney) 210
Sunday Sun 222–23
Sunday Sun & Guardian 207
Sunday Telegraph (Sydney) 192–93, 198
supplies (EF). *See* equipment (EF)
Supply, Department of (Aus.) 29–30, 53, 242, 245
Supply, Ministry of (UK)
    and British atomic weapons program 48, 76, 88–89
    and EF 22
    Lindemann, correspondence with 94
    OT 21, 155–56
    Penney, meeting with 53
    RAAF and 182–83
Supply and Development Act 1939–1948 35

suppression of media reports 52
Sydney Morning Herald 199–200, 201–204, 208, 209–210
Symonds, John 183, 238

tanks 36, 54, 97–98, 240
target response experiments 54, 96–98, 193
Task Totem Alpha 169
technology, weapons 63–64
test sites, atomic. *See* atomic test sites
Theaker, Martin 65
thermonuclear weapons 63–64, 79, 218, 258
Titterton, Ernest
    on Aboriginal people 132
    on atomic tests in Australia 146
    on Australian observers 147
    and AWTSC 145
    correspondence: Adams 113; Brown 30; Penney 148–49; Redmond 39
    EF experiments 30
    Elmhirst on 107–108
    Lester, Y, infuriated by 267
    on McClellan 267
    McClelland on 30
    McKnight on 32–33, 148
    OH 31
    OT: information received 154; involvement 30–33, 98, 107–110, 146–47, 195, 208; safety evaluation 147–50
    Penney and 31, 32, 145, 263
    RC 146, 264–67
    Stevens, JES, and 108
Tom-Cat, Operation. *See* Operation Totem (EF)
Tonkin, David 131
Toon, Erron (Terry) 157–58
Totem I (EF)
    atomic cloud 95, 96, 102, 111, 112, 120–21, 127, 164–65, 170, 174–77, 182–83, 188, 201, 204
    bomb tower 13, 57, 93, 129

design 93
detonation 92, 102, 164, 174, 176, 186, 200–202, 204
explosive yield 58, 93, 103, 107, 145, 152–53
fallout 94, 115, 125
ground zero 13, 102–104, 268
heat 12, 106, 107
lethal range 109
meteorology 12, 91, 94–95, 120–21, 144–45
radiation exposure 98, 104–105, 164–65, 167–68, 170, 172, 176–78, 180, 183–84
radiation monitoring 104–106, 108, 167–68, 183–84, 185–86
radioactive contamination 95, 105, 111, 121, 125–30, 167–70, 171–79, 180, 183–84, 268–70
rockets 102–103
secrecy 202, 210
shock wave 96
warnings against detonation 12, 120, 144–45
Totem II (EF)
atomic cloud 95–96, 111, 112, 115
bomb tower 13, 57, 93
design 93
detonation 115
explosive yield 57–58, 92–93, 107, 145
fallout 94–95, 115, 238
ground zero 13, 268
heat 12, 107
lethal range 110
meteorology 94–95, 114, 238
radiation exposure 173
radioactive contamination 95, 110, 114, 168, 173–74, 238, 268–70
rehearsals 87
rockets 102–103, 178
secrecy 92, 112–13
shock wave 96
Totem Alpha, Task 169

Totem Beta 170, 171
Totem Executive (TOTEX) 22, 51–52
Totem Gamma 170
Totem Panel (Aus.)
 on EF construction 53–55
 on Jindiviks 165–66
 and media 53, 55, 191–92
 members 61
 RAAF representative 169
 on safety evaluations 148
 and Woomera Prohibited Area 34–35
TOTEX (Totem Executive, UK) 22, 51–52
tourists 101
Townley, Athol 272–74
Townsville 95, 111
Treaty on the Non-Proliferation of Nuclear Weapons 235
trees 1–2, 269
trinitite 13, 102, 262, 276–77
Truman, Harry S 75, 77
truth, real and perceived 137–40, 161–62
Tube Alloys 58
Tyte, LC 49

UK government. *See* British government
United Kingdom Atomic Energy Authority 89, 234
United Kingdom government. *See* British government
United Aborigines Mission, Ooldea 10
United Nations 230, 234, 235
United States government. *See* US government
United States Air Force 151, 165, 173, 174, 183–84
unpredictability, British weapons 71–72, 73, 79–80, 143
uranium
 Australian, negotiations over 209–210, 212–36

British 70–73
Commonwealth 83–84, 216–17, 218
uranium-235 70, 72
uranium-238 84
uranium ore 71
uranium oxide 64, 70
urchins (initiators) 84, 85
urgency 37, 47–49, 60–61, 93
US Air Force 151, 165, 173, 174, 183–84
US government
   Australia, delegation to 234
   British government and 17–18, 74–75, 78, 89–90
   media access to nuclear tests 189–90
   OT 207
   thermonuclear weapons 218
   See also McMahon Act (Atomic Energy Act 1946, US); nuclear co-operation; Quebec Agreement; uranium: Australian, negotiations over

V bomber aircraft 68, 69, 87
Vallis, DG 109
vehicles 36, 50, 243. *See also* aircraft
Vincent, Eve 8, 25
visitor guides 142, 153
visitor permits 6–7, 270

Walker, EW 99
Wallatinna (Wallatina) 7
   anxiety at 140
   black mist 12, 117, 126–27, 129
   black mist commemoration 15, 136
   eye problems at 133
   RC 14, 118–20, 137, 152
   See also Lester, Yami
Walsh, Peter 13
war, atomic 24, 81–82
waste, contaminated 50
water 8, 29–30, 45, 49, 161
Watson, Maxwell 169
weapons, atomic. *See* atomic weapons
weapons, nuclear. *See* atomic weapons; thermonuclear weapons
weapons, thermonuclear. *See* thermonuclear weapons
Weapons Research Establishment (Aus., *formerly* LRWE) 272
Welbourn Hill 121, 126, 135, 152
'Welcome to the Claypan' booklet 153–54
West Australian 271
White (flight lieutenant, UK) 86
White, Jack 216
Whitlam, Gough 15
Wilkens, PH 86
Williams, Geoff 126–29, 176
Williams, Noel 237, 251, 252–53
Wilmot, John 65, 73
Wilson, Denis 175, 176, 178, 179, 182
Wilson cloud 121–22
Windscale atomic piles 45–46
Wise, Keith 264
Woomera Prohibited Area 6–7, 23–24, 28–29, 34–35, 42
Woomera rocket testing range
   EF conflated with 196–97
   establishment 26
   media reports on 193–94
   OT and 23–24, 270–71
   permanency 238
   rocket tests 9, 23, 196, 271–74
   Sandys on 223
   Southwell on 238

X200. *See* Emu Field test site (X200)
X300. *See* Maralinga test site (X300)

yields. *See* explosive yields